T0074694

Machine Learning in Cognitive IoT

Machine Learning in Cognitive IoT

By

Neeraj Kumar
Aaisha Makkar

CRC Press is an imprint of the
Taylor & Francis Group, an **informa** business

CRC Press
Taylor & Francis Group
6000 Broken Sound Parkway NW, Suite 300
Boca Raton, FL 33487-2742

Printed on acid-free paper
International Standard Book Number-13: 978-0-367-35916-4 (Hardback)

Visit the Taylor & Francis Web site at
http://www.taylorandfrancis.com

and the CRC Press Web site at
http://www.crcpress.com

To my little princess, my loving brother, my husband and my dearest parents.

Contents

Preface

I am delighted to introduce my first book on Machine Learning in Cognitive IoT. When I came to know about this book proposal, I was intrigued by the topic and instantly wanted to work on it with my supervisor's guidance. I was pleased that this book is coming in the early stage of a field that will need it more than most fields do. In the most emerging research fields, this book can play a significant role in bringing some maturity to the field. This book proposal is being accepted with the objective to educate the new learners. I took this opportunity and aimed to provide knowledge that is diverse in nature and dedicated my time searching for the techniques being used in research communities. Well, to be very precise, with the help of little perspective obtained from research papers, I attempted to experiment with all related techniques in the adjoining topics. The application potential of the selected areas could be easily felt in today's cognitive era. This book can serve as the platform for its growth. I can assure that the information contained in this book could not be found in any other textbook. The agenda of the content is to bring together the disparate topics being searched for by the keen learners. The continuous guidance by my supervisor enabled me to collect brilliant material by research; completely dedicated to this book.

Machine learning started gaining recognition in the early 2000s as a predictive tool. Classification, regression, clustering, and association are the techniques given attention to by machine learning researchers to explore more advanced combinational techniques for better prediction. The role of machine learning with cognition sensing is an upcoming field which has been illustrated as real-life examples along with demonstrations in this book. The systems we are facilitated with the use of machine learning are in the form of one or more tools. The weather forecast and traveling time according to traffic analysis are a few examples of predictions done using machine learning.

Our lives are surrounded by smart devices that we can call IoT devices. These devices generate a large volume of data. Abstracting the essential information from this amount of data is a tedious job. Machine learning proposed in this book analyzes the data deeply and helps to summarize the data more precisely. Finding new data trends, analyzing business schemes, gathering the weather information, all this is done by exploring different datasets. The preprocessing of datasets to fetch data's hidden trends is discussed in this book. This book attempts to explicate the state-of-the-art research in cognitive ability in IoT using machine learning. The book is invaluable, topical, and timely and can serve nicely as a reference book for courses at both undergraduate and postgraduate levels. It can also serve as a key source of knowledge for scientists, professionals, researchers, and academicians, who are interested in new challenges, theories, practice, and advanced applications of machine learning. I am happy to inform the readers that *Machine Learning in Cognitive IoT* addresses important research directions in IoT and the development of machine learning models for analyzing the data generated by IoT devices. It marks an important step in the maturation of this field and will serve to unify, advance, and challenge the scientific community in many ways. I hope that the readers will find this book useful and a source of inspiration for their research and professional activities.

The Internet's transformation into the Internet of Things (IoT) is becoming the core of advanced devices. The interaction with the IoT devices and thus storing the essential information is a challenge. More like analyzing the gathered IoT data is also a valuable consideration. The information exchange and interaction among the IoT objects is vital for enabling seamless communication.

The machine learning is adopted widely for data analytics and as a prediction tool. Furthermore, it seems to predict the collaborated edge between IoT and machine learning. The connection between the two is for the beautification of data. This book targets to covers all the aspects of this collaboration. Cognitive science has broad horizons, which cover different characteristics of cognition. The field is highly trans-disciplinary in nature, combining ideas, principles, and methods of psychology, computer science, linguistics, philosophy, neuroscience, etc. Also, cognitive computing using machine learning is the creation of self-learning systems that use data mining to find hidden patterns without constant human oversight.

Cognitive computing will bring a high level of fluidity to data analytics. The chapters included in this book aimed at addressing recent trends, innovative ideas, challenges and solutions of cognitive computing with machine learning in IoT. Moreover, these chapters specify novel in-depth fundamental research contributions from a methodological/ application in data science accomplishing a sustainable solution for the future perspective. Further, this book provides a comprehensive overview of the constituent paradigms underlying machine learning methods, which are illustrating more attention to big data over IoT problems as they evolve. Hence, the main objective of the book is to facilitate a forum to a large variety of researchers, where machine learning under the IoT environment is adopted to demonstrate the usage of machine learning models in IoT data analysis.

Acknowledgements

My first real introduction to machine learning was in the course work of my PhD degree. This subject inspired me so that I decided to continue my research on it. The knowledge acquired from the various books, Internet and research papers, is uniformed with my keen interest. This book contains the code of each and every technique elaborated here, along with the output. This endeavor is completed by me with the full support and guidance of Dr. Neeraj Kumar, without whom it would not have been possible. My PhD research is based on updating the Google's ranking scheme, i.e., PageRank algorithm. Although, it is the major field of web mining but the potential of machine learning taught me to use it in my research. Using machine learning, I proposed a data cleaning approach, i.e., SOTU (split by over fitting and train by under fitting). It raised my confidence in machine learning and, finally, I accepted this book proposal.

This proposal is incomplete until I acknowledge the contributors who made it possible. Firstly I would like to thank God for granting me this opportunity and completing it without obstacles. I would like to express my gratitude towards Dr. Neeraj Kumar, who trusted my knowledge and assigned me with this task. He supported me throughout the complete journey with knowledge and patience. Words are often inadequate to express one's deepest regards, but let's give it a go. I would like to thank my family members, my husband, my mother, and my father for giving such space and time that I was able to complete this book within the resources. They have truly been the source of real inspiration for me, and I will always remain indebted to them. Apart from providing me with excellent supervision, strong cooperation, and constant encouragement throughout this journey, they also shared their invaluable experiences with me to succeed in life. I would also like to thank my friends and colleagues with whom I have traveled this journey of writing. A special thanks to my research group, all upcoming doctorates, Rajat Chaudhary, Ishan Budhiraja, Himanshu Sharma, Shubhani Aggarwal, and Arzoo Miglani. This team helped me with proofreading the content of this book.

List of Figures

List of Tables

1

Internet of Things

Just imagine the life without electricity, Internet, and home appliances. Today's modern conveniences acquired by end-users are due to the endless efforts of intelligent researchers across the globe. These scientists have made our life easy and comfortable. For example, let's start with daily life that deals with different types of tasks and activities. The time we wake up and till we get ready for our work, we use various gadgets like mobile chargers, microwaves, hairdryers, and espresso makers, to name a few. All these gadgets were developed to reduce human efforts and to save time. Similarly, various Internet-enabled smart devices such as computers, smartphones, and printers are connected together in order to improve workplace operations as the Internet enables intelligent communication and control by end-users anywhere, anytime via wired or wireless connection. Hence, we conclude that the Internet plays an important role in our daily life. This interconnection of smart devices with the Internet is termed the Internet of Things (IoT) as summarized in Figure 1.1. IoT can be defined as an interconnected system where every device which is addressable in the digitized world is connected using the Internet. IoT is defined by Gartner, as a 'Network of physical object that contains embedded technology to communicate and sense or interact with their internal states or to the external environment'. IoT devices not only make our tasks effortless but reduce the delay in executing these tasks. In today's competitive world, these tasks are mainly accomplished with smart devices. Smart devices usually have built-in sensors and API's used for communication purposes. Some of the popular use-cases of IoT sensors are summarized as follows.

- Smart Traffic Lights: The traffic lights use sensors to actuate the volume of traffic at a particular instant. This observation helps to adjust the timers to control the traffic light operations. The objective of such a system is to minimize traffic jams by granting priority to peak volume areas and synchronizing the others.
- Smart Plugs: These plugs are attached with the power hub of the office. They monitor energy usage by disabling the high electricity-consuming devices when not in use. These devices save money as well as the environment by reducing strain.

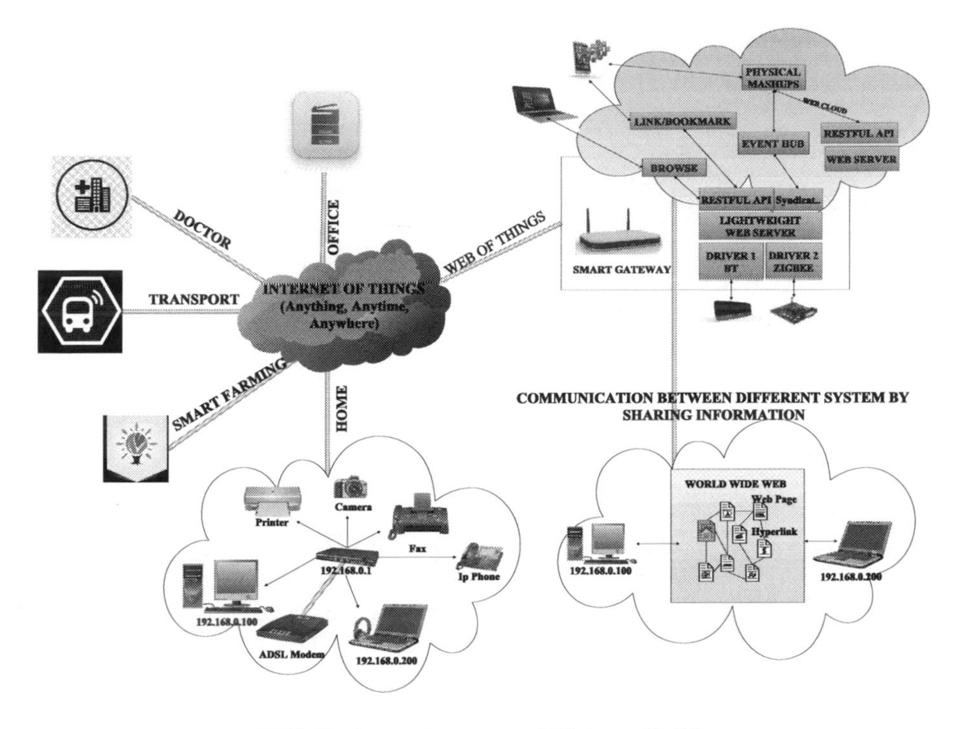

FIGURE 1.1: Internet of Things (IoT)

- Smart Mattresses: The smart sleepers have embedded sensors which tracks the sleeping position of a human. It notifies about the sleeping position and body pressure during the sleep to the human.
- Smart Blinds: These are the shades which protect the room from sunlight, by lowering the usage of energy. These blinds helps to maintain the temperature of a room. These blinds are easy to install and can be controlled by the smart phone.
- Smart Trash Cans: These cans provide the way to intelligently trash waste for maintaining the healthy home. It has sensors which sense the environment and control the opening/closing of its lid by measuring the fill level of the trash bin. It removes the germs and helps in cleaning the trash as well.
- Smart Shoes: These are the first insoles which heat the feet. They can also be controlled via smartphone. They monitors the distance covered during a day and amount of calories burned. They also helps prevent cold feet as they maintain the temperature of your footwear.
- Smart Helmets: Whether you're riding a road bike, mountain bike, ebike or motorcycle, it doesn't matter whether a good helmet is vital to staying

safe. But even the simple helmet is getting a high-tech makeover in this day and age, with all sorts of clever elements being added.

These devices are popularly known as IoT devices. IoT connects the devices for sharing endless information which can be summarized as 'connection', whereas embedding the intelligence into the devices is the task of artificial intelligence (AI). AI is that branch of computer science which has many sub-branches like machine learning (ML), natural language processing (NLP), deep learning, and many more. It can be defined by the term 'intelligence' as it is a program which handles the decision making without human interference. In summary, connecting the devices is the task of IoT, whereas putting the intelligence into it is done by AI. AI is not IoT, but IoT is what by which AI is incomplete.

1.1 IoT History

In recent years, there has been an exponential increase in the usage of the Internet. It is widely used for information retrieval. This information is gathered, stored, and processed at the central repository known as a web server. The server represents this information in the form of web documents, which are accessible with the help of the Internet. According to the *Statista* report, a web information provider company, the number of Internet users in 2018 was 369.01 million. The platform used for accessing the Internet is mostly the dedicated designed software known as a web search engine. According to *NETMARKETSHARE*, a market share statistics provider, the largest market share of search engines is achieved by Google, i.e., 72.03%, followed by Baidu(14.11%), Bing(7.76%) and Yahoo(4.27%), accessed on September 2018. The reason behind the success of web search engines are the search engine result pages (SERPs). These pages are ranked by the ranking methodology which considers the important features of a web page. PageRank is the ranking algorithm used by Google for ranking the web pages for SERPs.

Internet gained popularity in the 1980s, whereas Internet of Things (IoT) gained popularity in 2010s. The term IoT was coined by Kevin Ashton in 1999 at Procter in & Gamble. Internet of Network (IoN), Internet of Computers (IoC), and Internet of People (IoP) are the different phases of the Internet before IoT. IoT uses sensors and actuators in its devices to sense the physical environment. An overall mechanism of IoT connections is based upon the infrastructure of the Internet. The physical environment monitoring is done to take advantage of the virtual environment for information retrieval. This interconnection helps to capture the information-centric applications and services. The complete world today depends on IoT. A report from *Business Insider* states that the usage of IoT devices has increased from 14.2 billion in 2015 to 24 billion in 2019.

1.2 IoT Architecture

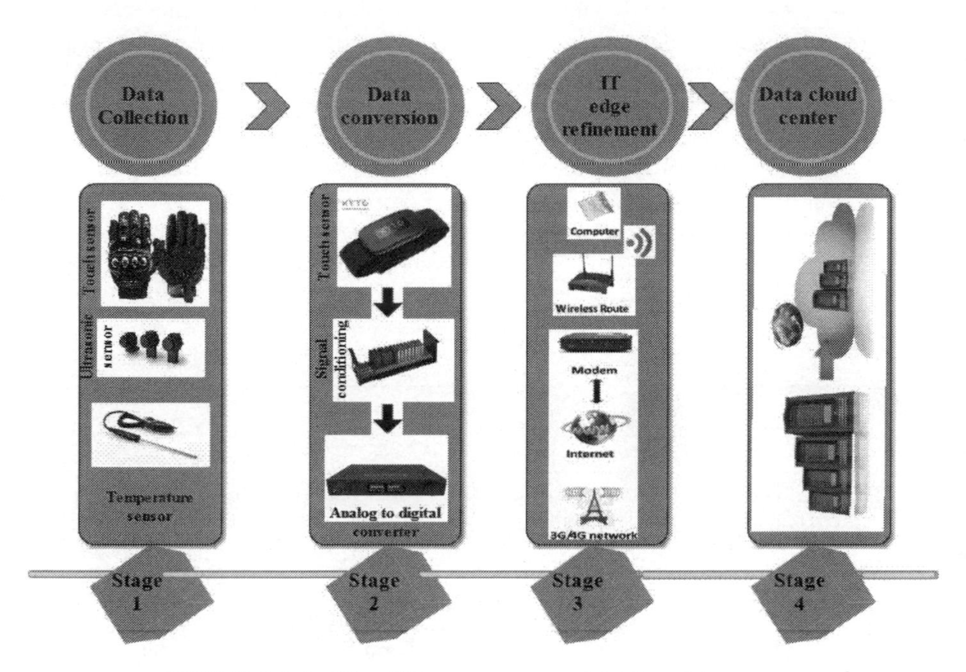

FIGURE 1.2: The different stages of IoT architecture

Can you imagine a place where the people are just sitting and receive the things they just imagine or wish to have? It is only possible when the lifeless things start thinking. Yes, when the things/devices are structured in such an architecture, the output is produced with 'brain' intelligence. The IoT architecture mostly depends upon the industry. It is the four-step process as shown in Figure 1.2. Integrated system architecture uses wired and wireless components. The effectiveness of such a system depends upon the quality of its components. These components include sensors, actuators, edge IT centers, data centers, and many other simulating components. The step-by-step process is required for the establishment of IoT architecture as discussed below:

- 1st Stage: Data collection
  The sensors collect the data by sensing the environment and convert it into useful information to be processed. They monitor the surroundings of the *thing* in which they are embedded. For example, the sensors on the fit band monitor the heart rate. The actuator acts like an automated part, which performs the immediate action to be taken. For example, shutting the power supply, adjusting the flow of air, changing the room temperature or so. You can imagine a fire extinguisher installed in a computer

lab when the room temperature gets heated and its temperature rises suddenly. The sensor installed on the fire extinguisher senses the temperature and in response the actuators blow an alert alarm. This is how the data/information is collected by the sensor, and immediate action is performed by an actuator.

- 2nd Stage: Data conversion
  The data collected by the sensor is in analog form which has to be converted into digital streams for further processing. The aggregation and conversion of data are done by data acquisition systems (DAS). DAS collects the data from the sensor, performs the analog-to-digital conversion and handles the data to the next stage. The Internet is the gateway used for this transmission. It can be viewed that this is the stage of data preprocessing. The analog data collected by the sensor is large in volume. The tasks performed at this stage include malware protection, data cleaning, data analysis, data management, and data filtering. Thus, it results in data acquisition and refinement.
- 3rd Stage: IT edge refinement
  Once the data is aggregated by the previous stage, it is ready for further refinement. This refinement process may or may not occur near the place where the sensors are installed. This stage has less burden of data transmission and reduces the load requirement for it. Congestion and other security issues are resolved at this stage. The anomalies are also detected by using machine learning. This is the final preprocessing of data before submitting it to the data center.
- 4th stage: Data cloud center
  This is the stage where the preprocessed data is analyzed more deeply, managed, and thus, processed. Two modules of this layer are operations technology (OT) and information technology (IT). The output can take time because of the involvement of various technologies. Collaboration and integration are the processes of OT and IT.

1.3 IoT Elements

1.3.1 Wireless Sensor Networks (WSN)

In today's era, remote sensing applications use advanced techniques, such as integrated circuits, which result in low power consumption and low cost. These factors have made the installment of sensor networks more convenient. A wireless sensor network (WSN) can sense, compute, process, communicate, analyze, and actuate. A sensor node transfers the collected data to another

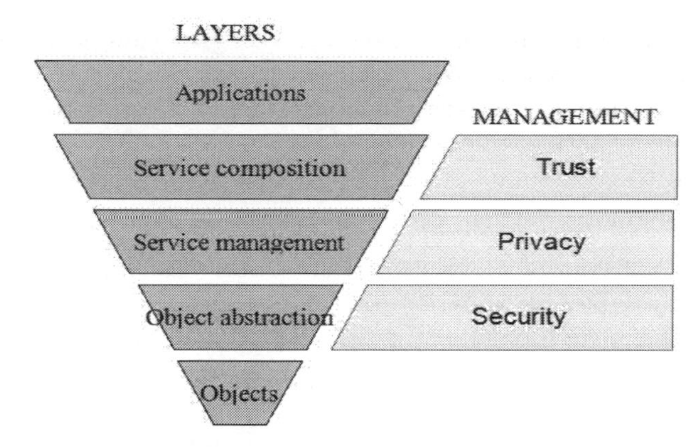

FIGURE 1.3: SOA architecture

sensor node by monitoring the environment using multi-hop transmission. Multi-hop routing helps in the transmission of data, and then the gateway helps the data to reach the management node. The sensor node is one of the integral parts of a WSN. It is the job of a sensor node to sense the environment. The sensor node then transfers the collected data to the microcontroller. The major components of the WSN network include:

- WSN hardware: It includes a power management module, sensor interfaces, transceiver, and processing units.
- WSN communication stack: This offers topology selection and routing mechanisms in such a way that it guarantees the longevity and efficiency of the network.
- WSN middleware: It is the software architecture used for expanding web-based services, i.e., Service Oriented Architecture (SOA) as shown in Figure 1.3.
- Security: The secure methods are required to heal the capacity of the network as network topology failure or sensor disability is very common. This security can keep intruders away from the network.

1.3.2 Radio Frequency Identification (RFID)

RFID technology is the wireless technology which helps in designing microchips for automated communication. RFID acts as an electronic bar code for the automatic identification of the objects to which it is attached. There is an RFID reader for communication with the RFID tags. RFID tags use the battery of an RFID reader. There is a particular range of operations from where the RFID systems are installed, and it varies from 100 kHz to 5.8 GHz. In other words, we can say it works in an Industry, Scientific, and Medical (ISM) frequency band.

The RFID tags use radio frequency (RF) electromagnetic energy. Sensor technology devices and sensor-capable devices embed RFID tags. These tags can sense, monitor, and record environmental changes. There are three different types of RFID tags as discussed below. The difference between these tags is listed in Table 1.1.

1. **Active tags**: These tags have batteries which send the radio frequency waves to RFID readers. They provide the capability for readers to detect the objects up to a range of 750 feet. These tags are always active and can receive the signals from the reader at any time. They can monitor as well as record the data (read with sensor). These tags are expensive and need maintenance due to the battery installation.
2. **Passive tags**: These tags do not have batteries and use the battery of RFID reader. These tags are very tiny and can be easily installed. They can only read the data and cannot record the data. They need good signal strength from the reader and should be placed close to the reader. These tags are very inexpensive and do not need much maintenance.
3. **Semi-passive tags**: These tags have a small battery installed on them. These tags have a good range of signals. They do not transmit active signals.

TABLE 1.1: Differences between different RFID tags

Difference	Active RFID	Passive RFID	Semi passive RFID
Internal Battery	Yes	No	Yes
Availability	Always	Within the range of reader	Within the range of reader
Cost	Most expensive	Less expensive	Lesser expensive
Capability	Always active to monitor and record the data	Active for monitoring but within the range of RFID	Active only when receives signal from reader
Signal Strength	High	low	Moderate
Range	More than hundred meters	Few meters	Around hundred meters

1.3.2.1 RFID Applications

RFID tags are popular nowadays. They can be seen in gadgets which are used in daily activities. Some of these devices are:

1. Transportation (replacement of tickets, registration stickers)

2. Automobile ignition keys
3. Bank cards (credit cards)
4. Road toll tags
5. Passports
6. Driving licenses (in the U.S., with electronic product code)
7. Anti-collision protocols
8. Aloha-based systems (pure and slotted aloha)
9. Bitwise arbitration algorithms
10. Binary search algorithms
11. Virtual and digital healthcare
12. Intervehicular communication
13. Oil and gas industry
14. Roads and bridges
15. Real-time location systems like Global Positioning Systems (GPS)

1.3.3 Data Storage

The surprising outcome of IoT architecture is the collection of a miraculous amount of data. There are some challenging issues considered while storing the data such as security, privacy, and confidentiality. The data update is one of the continuous processes carried throughout all the phases. The data analytics are performed mainly at the two phases of IoT architecture, i.e., third phase and fourth phase. To retrieve a meaningful sense from the data collected in the first and second stages, data analytics plays a crucial role. Specifically, data analytics remove the noise from the data. Many optimizations and fusion algorithms are developed to perform data analysis in IoT. Industrial IoT also makes use of data analytics to perform maintenance and efficiency operations. Data analytics is important in IoT for various roles, i.e., decision making, operations efficiency, competitive edge, and preventive maintenance. These roles are performed by various techniques such as ML, AI, and NLP. These techniques ensure communication among the various sources of data.

All the processed data is stored at the data center. These centers run on the harvested energy and are constructed in such a manner that they maximize data efficiency and reliability. This infrastructure is designed with a view of IT applications. Clearly, the support provided by the software and hardware becomes an integral part of the success of the system. This is how the data is gathered, processed, analyzed, and stored in the IoT infrastructure.

1.3.4 Challenging Issues

IoT 'Things', which we call devices or gadgets, should be uniquely identifiable. The success of the IoT system depends upon this constraint as well. For example, the request for data is made from a particular city, but it gets delivered to another city, who will you blame? It is the delivery system. So, to avoid this type of conflict, all the billions of devices should have a unique identity. The IoT devices are controlled by the Internet as we already discussed previously. There are few parameters to be kept in mind while designing the unique address, i.e., persistence, uniqueness, and scalability. There are the protocols which ensure reliable identification geographically but may not individually, such as IPv4. The TCP/IP is used for handling this routing behavior, but the wireless sensor networks and its gateway needs attention. The network of the 'things' should be configured in such a manner that the addition or removal of one should not affect the other. To ensure the reliability of the network, the Uniform Resource Name (URN) becomes an integral part of IoT structure configuration. URLs are created by URN for accessing the proper resources. The database maintains the constraints of data integrity and data confidentially. However, the metadata generated by the database is mainly adopted for passing the information over the Internet.

1.4 Data Analytics

After entering into the IoT architecture as an input, the data gets back to the user in the form of output. What happens to the data in the four stages of IoT architecture? How it is processed so that we get refined information? The complete process from data input to meaningful output is known as data analytics. This process is carried out by various activities, namely, IoT data sources, data preprocessing, technology selection, and optimization. All these activities are presented in Figure 1.2. The complete working of these activities is elaborated in Figure 1.4 and discussed below.

1.4.1 IoT Data Sources

The data entered by the applications, users, devices, and sensors are fetched by the IoT infrastructure. IoT processes the data in such a manner that the meaningful information is retrieved, which in turns help initiate the smart activity. IoT data can be generated by any enterprise/organization/device by using IoT service/application. Based on the IoT application deployed, the IoT data sources are categorized as listed as follows.

1.4.1.1 Industrial Data

A large amount of data produced by industry is gathered and stored using IoT. Irrespective of the industry type, the data can be easily handled by the advanced technique, i.e., ML. Next, data analytics is performed, and some meaningful prediction can be made. IoT has the potential of handling real-time data by applying cost-effective technology. The data is fetched by creating various types of views. It maximizes productivity and improves the efficiency of the industry.

1.4.1.2 Business Applications

There are business applications like Customer Relationship Management (CRM), Enterprise Asset Management (EAM), Enterprise Resource Planning (ERP), and Product Lifecycle Management (PLM), which collaborate with IoT technology. These applications are the smarter way to handle enterprise data. Customers can check their products or service updates from their remote location, and the technicians can directly approach the fault region without involving the business organization. It thus reduces maintenance costs and improves customer experiences.

1.4.1.3 Sensors and Devices

The sensors and actuators deployed by IoT devices sense the environmental data and send it for processing. The common sensors used are temperature, flow, pressure, and humidity sensors. Wearable devices like fitness bands and smartwatches are the most suitable examples to detect environment changes monitored by IoT.

1.4.1.4 Smartphones

Smartphones are the medium to access real-time media from anywhere. Even while traveling, the real time photos can be downloaded or uploaded to the Google drive or any social networking website. Smart applications like GPS and navigating and storing on the cloud have made the life of humans easier and comfortable. Services such as- iCloud and Google Drive uses IoT technology for storage.

1.4.2 Data Processing

The data, which may be gathered from different sources, is then processed and analyzed for making intelligent decisions. This process involves converting, filtering, and cleaning the data before it is transferred to the next module.

1.4.2.1 Data Acquisition

This is the process of converting the analog signals collected by the sensors into the digital signals. Analog signals can be defined as a continuous signal which changes its frequency over time. These signals are generated by the environment/human beings. Digital signals are the signals understandable by the computer. Analog-to-Digital Converter (ADC) is the hardware piece used for this purpose.

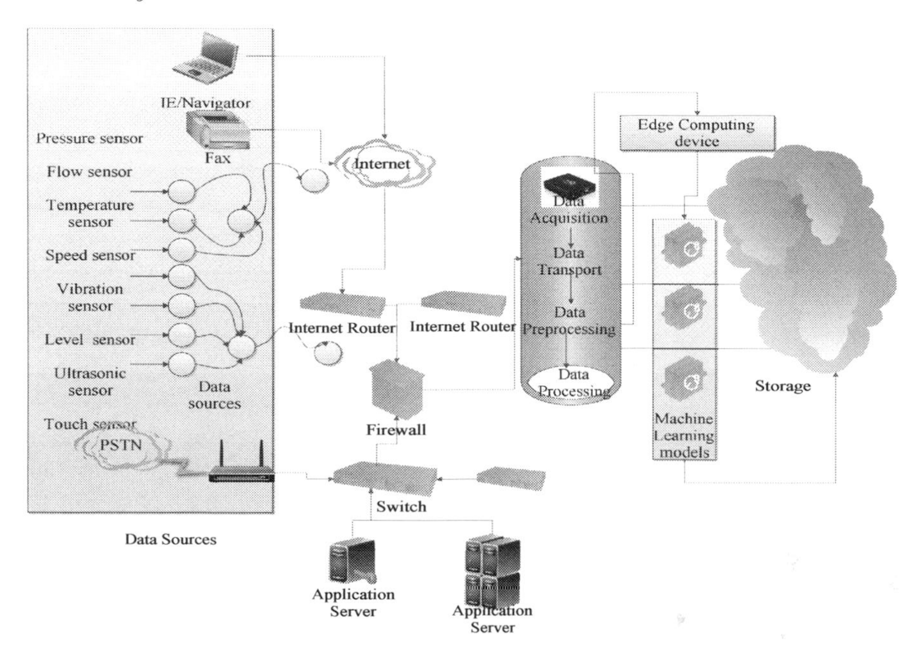

FIGURE 1.4: Overview of data analytics process from input to the final output

1.4.2.2 Data Transport

The data, depending upon its format, is gathered together and transported from the various repositories. The encryption techniques, if necessary, can be applied for secure data transport from source to destination. At this stage, the information spread during the conversion is collected at one place.

1.4.2.3 Data Preprocessing

The data prepared at the previous stage is in the digital form, ready to be processed. But, Is the data clean and sufficient enough to be fed into the technique? The data should be in consistent form. The data should not have missing values. All the doubts have to be resolved before experimenting with the data. There are various steps involved. To understand each step more clearly, data preprocessing is discussed in the next section.

1.4.3 IoT Technologies

The strategy adopted by the technology is in the context of ubiquitous wireless communication, ML, and data capturing elements, like embedded systems. The wireless techniques not only help in transmission but also ensure security.

1.4.3.1 Wireless Communication Technologies

Wireless communication technologies are used for the transportation of data over the data. There are various protocols such as IPv6, UDP, and many more as discussed in Section 1.5.

1.4.3.2 Data Analysis Schemes

Data analysis ensures that the data is ready to fit into the models of ML. The data at this stage is the level of computer responsiveness which flows from the processed data to make it more efficient. There are four main categories of data analytics described as follows:

1. **Descriptive**: It defines the past by gathering raw data from various data sources to extract useful information. The result describes what has happened but does not describe why it happened. For example, a student may gather the marks of all his classmates to judge who has scored the maximum.
2. **Diagnostic**: It is the process to analyze the reason behind some event which has already occurred. It gives us the answer to why it has happened. For example, a student finds the reason why he had scored poorer marks.
3. **Predictive**: It is the technique of forecasting the future by considering the past instances. It is done to know what will happen. It considers various techniques like machine learning, which uses various models for prediction. For example, forecasting the temperature of another city after two days for a visit.
4. **Prescriptive**: It is the method to learn from the experiences to avoid the near future risk. It gives us the answer as to what should we do. It involves various techniques. For example, a student learns new techniques to answer the questions in an examination to achieve good marks.

1.4.3.3 Machine Learning

The data which is processed previously is analyzed using the ML technique. This technique exploits the hidden insights of data and predicts the new data. There are various ML models like classification, regression, and clustering models. We train the processed data into one of these models, and then the model predicts the unseen data by learning from the previous data. This technique helps us to avoid failure and losses. For example, predicting tomorrow's temperature can be done using ML.

1.4.4 Optimization Techniques

Finally, we obtainthe results from the technique, i.e., ML is ready to become output after optimization. The optimized techniques help to improve the results.

1.4.4.1 Edge Computing Technology

Edge computing performs an optimization by integrating the intelligence, processing power, and communication capabilities near to the edge gateway or appliance directly into devices like programmable automation controllers. It results in low latency, minimum bandwidth consumption, and data integrity.

1.4.4.2 Refining the Results

Finally, the output is produced by adopting the cross-validation and ensemble approach to ML. It improves the overall accuracy of the results.

1.5 Steps of Data Preprocessing

The data used by machine learning models should be clean, noise-free, sufficient, and consistent. The relevancy of data ensures the best output. There are various steps to be used in the data preprocessing.

1.5.1 Data Formatting

This step refers to the procedure required for formatting the data. As the data is collected from various sources and then converted by ADC, the format should be the same and consistent for all the data. For example, when you are practicing machine learning and download the dataset from Google, you may observe different types of files being downloaded, such as, .txt, .pl, .graph, .csv. All the data needs to be formatted before it is fed to the machine learning models. For example, using software like R or RStudio requires a .csv file for the processing. So the data needs to be formatted in the .csv file.

1.5.2 Data Cleaning

The data should be clean enough for further processing. Data cleaning is the crucial step of data preprocessing that ensures the detection of duplicate values, irrelevant values, and missing values. All these issues should be resolved, otherwise, it leads to the failure of machine learning models.

1.5.2.1 Duplicate Observations

Sometimes, the IoT sensors capture the same data again and again. For example, if you are leaving your child sleeping at home under CCTV footage and you leave for office. After returning home hours later, it won't be possible to see the complete footage of hours. The CCTV should be smart enough so that it should record only when a child moves, and not in the sleeping position. So, repetition of the same action should not be recorded. hence saving cost and time.

1.5.2.2 Irrelevant Observations

Collected data should not contain irrelevant information. For example, suppose you download the dataset of websites, and it contains information about the color of the website, which is irrelevant. It is best to detect such irrelevant observations.

1.5.2.3 Handling Missing Data

Sometimes, the data is in free form, and the values may consist of some zeros or null. It is very important to treat such values before the data is further processed. There are various ways to handle this constraint. A few of these are described as follows.

- Deleting Rows: Delete the complete column which contains approximately 70% to 90% of zero values.
- Replacing with Mean/Median/Mode: Substitute the missing value or the null value with some aggregate value, such as the mean.

1.5.2.4 Detecting Outliers

Outliers are the instances which does not lie within a particular range. It is difficult to detect the outliers just by visualization. It depends upon the data and its features as to how relevant and useful the data is.

1.5.3 Data Reduction Schemes

Real-time data is always collected randomly. As in the analog form, it is difficult to say the number of classes it will form or the number of dimensions it will converge to after conversion. So the resulting digital form of data may contain an unbalanced number of cases in various dimensions. Balancing the data is done under this step of data preprocessing. Dimensionality reduction is one of the major steps toward significant data abstraction. Dimensionality reduction is achievable in two stages, i.e., feature extraction and feature selection. Feature extraction is the approach used to reduce the high-dimensional features into low-dimensional features by forming the new feature space. However, feature extraction does not reduce the number of features, but the complexity of the features. Feature selection aims at removing the irrelevant and redundant features by forming a small set of features. These two techniques are elaborated below.

1.5.3.1 Feature Extraction

Feature extraction is one of the dimensionality reduction techniques, which reduces the number of resources used to describe the data. This technique reduces issues like overfitting, large memory requirements, and computation power. There are various feature extraction techniques, such as independent component analysis, principal component analysis, multifactor dimensionality reduction, nonlinear dimensionality reduction, multilinear, and many more. Above all, principal component analysis (PCA) is the most adopted. PCA is a multivariate procedure which transforms the box of correlated features into the box of uncorrelated features. The new set is formed in such a way that the maximum variance is extracted from the variables. The extracted variance is removed, thus forming the second combination of variables. This process is continued until the correlation between the variables is diminished. It results in orthogonal (uncorrelated) variance and the procedure is known as the principal axis method. It reduces the dimension of feature space. PCA is also known as a data compression technique. It reduces the complexity of data.

PCA computes the principal component (PC) scores by the multiplication of the standardized value of each row with the multiplication of the standardized value of each column. The nature of PCs which are computed depends upon the method adopted to compute PCA. There are different methods for computing PCA, and it also depends upon the software you are using. For example, R is adopted to perform the experiments, then the methods used for computing PCA are prcomp(), primcomp(), and PCA().

1.5.3.2 Feature Selection

The feature extraction reduces the dimension of the data, not the number of features. To eliminate the irrelevant features, there are many feature selection techniques as presented in Figure 1.5. The different categories of feature selection are briefly discussed below and detailed description is given in Chapter 8.

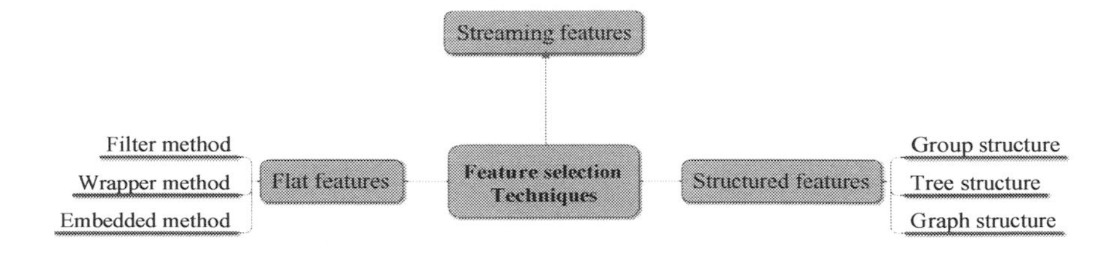

FIGURE 1.5: Different feature selection techniques

1. Flat features

    - Filter method: This method computes the correlation between the features and eliminates redundant features.
    - Wrapper method: This method is implemented using a machine learning model. By continuously observing the results, the combination of various features has experimented.
    - Embedded method: It is the method in which machine learning model itself selects the features, such as LASSO and RIDGE regression.

2. Streaming features: It is the method of sequentially adding the feature in a predictive model which reduces the subspace.
3. Structured features: The selection of features is done by considering some structure of features such as a group, tree, or graph.

1.6 IoT Protocols

There are many protocols used in different layers and stages of IoT architecture. Many Internet protocols are also used in IoT, but the IoT protocols reduce the overhead of parsing and overhead generation during data transmission also. To understand the protocols more precisely, all the protocols are categorized into the layers in which they work as illustrated in Table. 1.2. The underlying differences between the different protocols used is discussed in Table. 1.3.

TABLE 1.2: Different protocols used in IoT

S. No.	**Layer**	**Protocols**
1	Infrastructure (Network /Transport)	IPv6, 6LoWPAN, UDP, QUIC, uIP
2	Data	MQTT, MQTT-SN, Mosquitto, CoAP
3	Physical	Cellular (GPRS/2G/3G/4G/5G), RFID, Wi-Fi, Wireless HART, Zigbee, Bluetooth Low Energy
4	LPWAN	NB-IoT, LTE-MTC, EC-GSM-IoT, LoRaWAN, RPMA

1.6.1 Infrastructure Layer (Network/Transport Layer)

1. IPv4/IPv6: Both of these are the Internet protocols used for end-to-end transmission to multiple IP addresses. IPv4 was developed in 1980 whereas

IPv6 came in 1998. However, due to an increasing number of Internet devices, IPv6 replaced IPv4. The IPv6 is adopted by IoT due to security, scalability, and connectability.

2. 6LoWPAN: This protocol carries the data in the form of IPv6 in Wireless Sensor Networks (WSN). That is why it is popularly named as IPv6 over Low Power Wireless Personal Area Networks. It is built on the top of IEEE 802.15.4 with the specific frequency range (2.4 GHz) and specific transfer rate (250 kbps). But this need is now eliminated, and it can work easily. It can be installed on personal devices which can be operated by smartphones, such as, LED.

3. User Datagram Protocol (UDP): It works like TCP which sends datagrams for process-to-process communication. Port numbers' assignment and checksum computation are the two major advantages of UDP over TCP.

4. Quick UDP Internet Connections (QUIC): This was built to be used on the top of UDP and reduces the end-to-end latency. QUIC was adopted by IoT because of faster connection establishment, multiplexing, forward error correction (FEC), authentication and encryption, and better signaling.

5. uIP: It is popularly known as micro IP to be fit in the 8-bit or 16-bit microcontroller. It uses the memory of one packet buffer and is used to keep track of IP packets.

1.6.2 Data Protocols

1. Message Queuing Telemetry Transport (MQTT): This is used for lightweight message transport. It was developed for machine-to-machine and IoT communication. MQTT-SN is the MQTT for sensor networks which are widely used in IoT. It is useful at the places which have premium network bandwidth.

2. Mosquitto: The implementation of MQTT is achieved using Mosquitto which is designed for this purpose. It is suitable for low-power sensor devices such as smartphones (IoT device).

3. Constrained Application Protocol (CoAP): This is the web transfer protocol with the low RAM and ROM, specially designed for lower-power applications. It works somehow similar to 6LoWPAN.

1.6.3 Physical Layer

1. Cellular (GPRS/2G/3G/4Gs/5G): The technique used for longer distances and high-power rate. It is used in technologies where there is high-speed connectivity to the Internet. As this technology is widely used in smartphones, it is updated frequently.

2. RFID: Radio Frequency Identification was already discussed in detail (see Section 1.3.2).
3. Wi-Fi: This is based upon two IEEE standards, i.e., 802.11ah and 802.11ax. It is the technology used for transmitting data over the network over the long distances and at high speed.
4. Wireless HART: The wireless technology based on Highway Addressable Remote Transducer Protocol (HART). It works with IEEE 802.15.4 standards. This protocol uses mesh architecture.
5. ZigBee: The Zigbee protocol adopted the IEEE 802.15.4 standard. It works in the frequency range of 2.4 GHz with a speed of 250 kbps. It is used in IoT because of features such as connectivity and stability, reliable mesh network, and security.
6. Bluetooth: Bluetooth works in the 2.4 GHz ISM band and uses frequency hopping with a data rate up to 3 Mbps and a maximum range of 100 m.

1.6.4 LPWAN: Low Power Wide Area Network

1. NB-IoT: Narrowband IoT (NB-IoT) is LPWAN radio technology. It is adopted by IoT because of indoor coverage, long battery life, low cost, and high connection density.
2. LTE-MTC: This is the Long Term Evolution for Machines with the battery life of more than 10 years. It is supported by all the mobile devices, including all the cellular networks. It reduces the cost of currently enhanced EGPRS modems.
3. EC-GSM-IoT: This has given rise to existing cellular networks with few new capabilities. It can be easily installed with new software which covers a variety of IoT applications and serves more IoT devices.
4. LoRaWAN: This allows low-cost transmission over the wide-area range. It is designed in such a manner that it supports the applications and sensors to a small amount of information and has a long battery life.

1.7 IoT Applications

IoT devices are used in almost every activity of our daily routine. IoT has changed the world with emerging techniques and innovations. However, if we start discussing IoT applications, these are vast and endless. IoT applications and their major key elements are presented in Figure 1.6. Some of the major applications of IoT are listed in the following sections.

TABLE 1.3: Differences between protocols most used in IoT

Protocols	Range	Topology	Power	Data speed	Privacy	Frequency
NFC	10 cm	Point to point	Zero	Low	High	13.56 MHz
Bluetooth	10 m	Point to point	Low	high	Mid	2.4 GHz
Zigbee	10-20 m	Mesh	Low	High	Mid	2.4 GHz
Wi-Fi	30-100 m	Star	High	High	Low	2.4 GHz
LoWPAN	10-100 m	Star, Mesh	Low	Mid	High	2.4 GHz
RFID	0-200 m	Point to point	Low	Mid	Mid	902-928 MHz
Cellular	Several kms	NA	High	NA	Low	NA

1.7.1 Logistics and Transportation

Today, roads and vehicles are smart with the help of smart IoT innovations, such as smart traffic lights. Such devices help control traffic and avoid accidents. Even GPS technology has made the driver's journey easier. Vehicles also use sensors which control the vehicle with a remote. The driving learners should prefer to ride the cars equipped with sensors and actuators. Transporting goods from one place to another is also just a matter of a few clicks. The supply chain system of the enterprises can be managed by supplying goods at the convenience earliest and conveniently delivering them far-off places.

1.7.2 Home and Workplace

Home and workplace are now occupied with sensor- and actuator-embedded equipment, which help us comfortably perform routine activities such as heating food in the microwave, maintaining room temperature with air conditioning, storing vegetables and fruits for a longer time in the refrigerator, and a single command for printing documents. Such devices are designed for saving electricity by turning off the device when not in use.

1.7.3 Personal and Social

The World Wide Web, rightly called, is the most accessible network for being socially active online. Thanks to applications such as Facebook and Twitter, we connect to people and are aware of their activities. We may think of the sensor-based applications, which collects information about our visited place and updates us to pending places we need to visit according to the events in our social network. The search engine and the web play an important role in being active socially and personally. Other gadgets like fitbands help us to

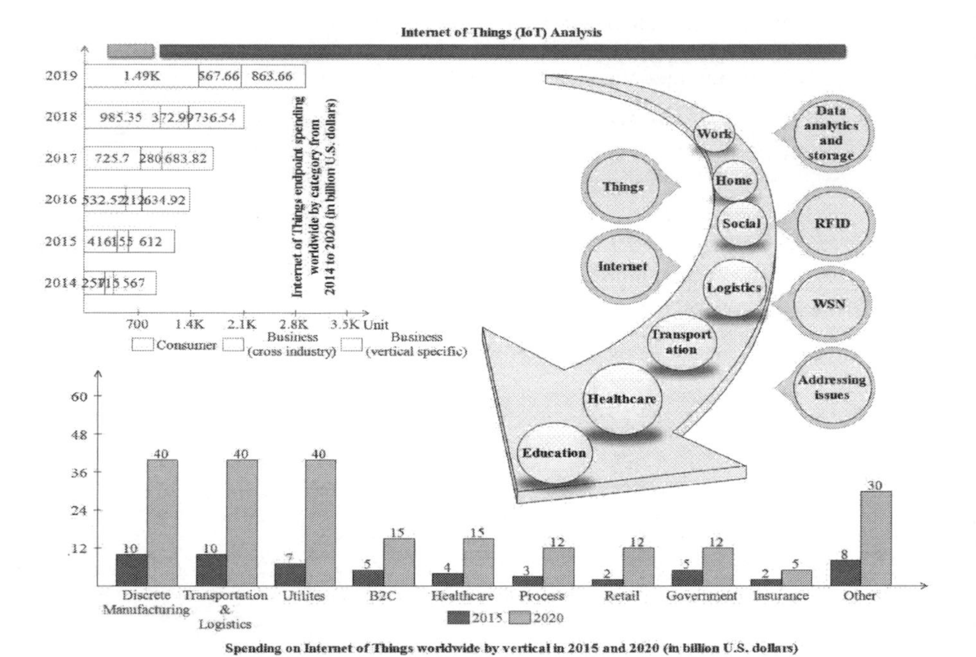

FIGURE 1.6: IoT applications and their trends over the years

monitor our health with embedded sensors which record footsteps, heart rate, and many more personal activities.

1.7.4 Health Domain

In the healthcare domain, IoT plays an important role. The medical diagnosis, health-monitoring types of equipment help to monitor and track patients' records. It helps identify and authenticate people, detecting various diseases and keeping the health track for years. Simultaneously, health monitoring can now also be done with the help of smartphone applications. Such data remains stored for a longer duration. Paper bills and their costs have also been reduced using online bill technology. Remote medical assistance is available nowadays in case of emergency from any remote location.

1.8 Book Outline

In our day-to-day activities, we have been using many electronic types of equipment. This equipment simplifies our daily life, but do they work and are they really smart? How we can make our home equipment smart and intelligent? How we can predict unseen circumstances? What are smart devices and how pervasively they are likely to get? There are large number of gadgets which help us in routine work. But, these gadgets are not so smart. In the cognitive era, we can make these gadgets more smartly to sense, reason, think, and react. These devices are developed in such a way that they meet their human counterparts up to an extent. How are these devices connected with the Internet, and what is the role of IoT in our daily life? How do machines, like computer systems, understand the human voices like in voice search of Google? How are the voice signals interpreted? These are the questions that come in the mind of users. This book tries to answer all such questions. A few more technical questions answered in this book include:

- What is IoT? How does it differ from AI?
- What are the IoT trends, and how have they changed human life?
- What are the components of cognitive behavior?
- How do cognitive things differ from IoT devices?
- What is the Cognitive Internet of Things (CIoT) and its techniques?
- What is the role of machine learning in CIoT?

This book covers the different related technologies to the Internet, and machine learning capabilities involved in CIoT. Machine learning is explored covering all the technical issues like models. Different categories of machine learning schemes are discussed.

The first chapter explores the Internet of Things (IoT), its history, architecture, and applications. The second chapter focuses on the concept of CIoT, and its capabilities in the real world. The third chapter describes the importance of machine learning in data mining for CIoT. The fourth chapter focuses on the various machine learning techniques, the difference between supervised learning and unsupervised learning, classification, clustering, and regression. The fifth chapter covers the paradigms of machine learning like classification rules, numeric prediction, and many more. The sixth chapter discusses the terminology used for the development of various machine learning models. The seventh chapter explores the different phases of data preprocessing like data cleaning. It also illustrates the programs of different data cleaning techniques that exist in literature like Tomek and SMOTE. The eighth chapter introduces feature engineering, i.e., feature extraction and feature selection. This chapter also gives brief insights for the optimization of results generated by a machine learning model. The last chapter covers the evaluation parameters for the validation of results. These parameters are such as accuracy, confusion matrix, roc and many more.

1.9 Target Audience

This book is primarily targeted for the learners, students, researchers, and industrialists, those who are planning to design an IoT infrastructure with emphasis on how machine learning can help the system. The cognition is the power to emphasize the upcoming facts of the technical world. So, cognitive computing is also an integral part of the system to make it smarter. This book provides much-needed information about potential IT strategies, IoT structure, cognitive computing, and machine learning. The book is written in such a way that it helps the developers and researchers to design customized solutions for different applications. It covers different types of machine learning models and their formation with numerical examples. It covers the details of machine learning terminology along with using R programs. Universities can use this book as reading material for their data mining/machine learning courses.

1.10 Summary and What's Next?

In this chapter, we have given attention to the need for IoT in today's era. It includes IoT history on how it is evolved from the Internet and discusses its capabilities. It is important to understand the architecture of IoT, which is widely used in the industry nowadays. The key elements readers need to understand before going into the insights of IoT are discussed. Data analytics play an important role as that is concerned with data processing activities such as data gathering, data preprocessing, and simulation technologies. Working on IoT devices and their applications are the reason behind the technical innovations. IoT has given rise to emerging technologies, i.e., CIoT is discussed in the next chapter. The key elements included in the book are as follows.

1. IoT and its advancements
2. CIoT and cognitive devices
3. How are IoT and CIoT dependent upon machine learning?
4. Machine learning techniques
5. Different machine learning models
6. Working of machine learning models
7. IoT, CIoT, and machine learning as a collaboration

1.11 Exercises

Questions

1. Define IoT? Differentiate between IoT and AI.
2. Why is IoT the most promising technology in exchanging information?
3. Describe the IoT architecture with its components.
4. Define the role of sensors in WSN with all their components.
5. Differentiate between RFID and WSM. Describe RFID in detail.
6. What is the need of data analytics in IoT?
7. What are the different IoT data sources?
8. What is the role of edge computing devices in data analytics process?
9. What are the different IoT applications?
10. What are the different protocols deployed at the network layer of IoT architecture?

Multiple Choice Questions

1: What is used for the aggregation and conversion of data in IoT architecture?

(a) Digitizers and Analog Systems (DAS) (b) Sensors

(c) Data Acquisition Systems (DAS) (d) Actuators

2: Which component of WSM is responsible for topology selection?

(a) WSN communication stack (b) WSN hardware

(c) WSN middleware (d) WSN security

3: From where does the RFID tag use a battery?

(a) Device on which it is installed
(b) RFID reader
(c) RFID tag
(d) ISM

4: Which RFID tag is most expensive?

(a) Active RFID tag
(b) Passive RFID tag
(c) Semi-passive RFID tag
(d) All of them

5: Which RFID tag should be installed in the car key lock?

(a) Active RFID tag
(b) Passive RFID tag
(c) Semi-passive RFID tag
(d) None of them

6: Which IoT component assigns the unique identification to the IoT devices?

(a) Internet
(b) Protocols
(c) WSN
(d) None of them

7: What is the full form of FEC?

(a) Forward Error Correction
(b) Fast Error Correction
(c) Forward Estimation Component
(d) Fast Estimation Correction

8: Which IoT protocol is responsible for routing behavior?

(a) Internet protocol
(b) TCP/IP
(c) URN
(d) None of them

9: Which IoT application is installed in a smartphone for navigation purpose?

(a) Smart route technology
(b) GPS technology
(c) iCloud technology
(d) None of above

10: Which hardware is used for analyzing digital signals?

(a) ADC
(b) Internet router
(c) Application server
(d) Machine learning

11: Which protocol sends datagrams for process-to-process communication?

(a) IPv4/IPv6
(b) 6LoWPAN
(c) UDP
(d) uIP

12: Which architecture is used by wireless HART protocol?

(a) Bus
(b) Star
(c) Mesh
(d) Hybrid

13: Which architecture is used by Wifi protocol?

(a) Bus
(b) Star
(c) Mesh
(d) Hybrid

14: Which is the role of LTE-MTC?

(a) Bus
(b) Star
(c) Mesh
(d) Hybrid

15: Which web transfer protocol is used for lower-power applications?

(a) MQTT
(b) Mosquitto
(c) CoAP
(d) TCP/IP

16: What is the transfer speed of Zigbee?

(a) 250 Kbps
(b) 750 Kbps
(c) 1 Mbps
(d) 3 Mbps

17: What are the different factors to be kept in mind while storing the IoT data?

(a) Security
(b) Privacy
(c) Confidentiality
(d) All of the above

18: Which data preprocessing step ensures data integrity?

(a) Data handling
(b) Data cleaning
(c) Detecting outliers
(d) Handling missing values

19: Which data analytic category defines the reason for the activity?

(a) Descriptive
(b) Diagnostic
(c) Predictive
(d) Prescriptive

20: Which technique helps reduce the dimensional of data?

(a) Feature selection
(b) Feature extraction
(c) Detecting outliers
(d) None of the above

2

Cognitive Internet of Things

The term *cognitive* means capable of being reduced to empirical factual knowledge. This statement itself states the reduction of believing only in facts as per the surroundings. The actual meaning of the term cognitive merely depends upon its participating agent. The cognitive term is used in various science fields—cognitive psychology, cognitive therapy, cognitive development, and cognitive impairment. Each field offers a different scenario for the use of this term. Cognitive psychology is the field of investigation for all the brain processes such as memory, attention, thinking, problem solving, and decision making. Cognitive therapy is the treatment given to the patients facing the symptoms of mental disorder such as anxiety, fear, depression and many more. Cognitive development is the field used for developing the thought processes that involve problem solving and decision making. Cognitive impairment is the field for improving memory-related issues such as long-term remembering, thinking, decision making, and concentrating. By learning from all the above definitions, we can say that the term 'cognitive' is related to memory processes in one sense or the other. Whereas artificial intelligence (AI) is the field of giving the intelligence power to the system artificially. However, the intelligence of the system is limited to what is coded in the system by humans. The success of AI systems depend upon the human's ability to code a solution for the logic and the scenarios for decision making. This is the constraint of AI, the roadblock to think beyond human thinking and the behavior response of humans. To get out of this, AI embraced the cognition ability and termed it as Cognitive AI. It is regarded as an extension to the existing AI paradigm that fills the gap between humans and machines. The cognitive computing is performed by various technologies such as machine learning, AI, NLP, and many more.

2.1 Cognitive Devices

The computing era has evolved over the last hundred years. The device carrying the cognition power, i.e., ability to think in terms of computer science, is popularly known as cognition device. The cognitive device can think, thus removing the prejudice of other devices. Cognitive devices/systems may learn from the environment, data, learning, and experience, whereas AI devices can perform only those tasks coded by humans. For instance, a human being learns from experience, so does the cognitive device. This auto-learning feature of the cognitive device makes it smarter, intelligent, and updated with the time and experience. You can imagine the forthcoming technology with this cognitive evolution. We talked about the IoT devices in the previous chapter. IoT devices are smart enough because of the sensing technology, but these devices are not intelligent. Embedding intelligence is the game of a cognition device. It thinks and acts like humans. You can take an example of an IoT device, like a smart shoe. It may count your footsteps but cannot differentiate the footsteps of a child and adult. Though, it can also count the calories burnt in a day but cannot instruct you for the same. Now, let us take an example of a cognitive device. Suppose you enter into the smart home, and the electric appliances get switched on. We can call it as hardcoded and automated. But if only the required appliances are turned on according to time, weather and need, it is soft coded and learns with the surroundings. This ability to learn, adapt, and react is the task of cognitive devices. This learning of a device not only performs the basic task but also the advanced tasks. Cognitive devices learn from their experiences, and they are soft coded. The other example is CCTV cameras are IoT devices, but if they record and then react as well, the system becomes cognitive.

2.2 Cognitive in IoT

The power of sensing and reacting in IoT restricts it from learning and remembering. Though, IoT devices reduce human efforts to some extent, due to the limited scale and features of IoT, as it is human coded, cognitive technology took devices to the next level. The handshake by the IoT with cognitive technology is popularly named Cognitive Internet of Things (CIoT). IoT devices with cognition power can be collectively called CIoT devices. They help in decision making, according to the situation. At the same time, it learns from the example, memorizes and then revises it if the same circumstances occur. The reasoning is applied in CIoT by analyzing the data for inferring useful information required for concluding a decision. Recent methodologies adopted by the CIoT technology are discussed as follows.

- Cognitive management framework was proposed to empower the IoT in smart cities for collaborative objects, by treating the real world as a virtual world. The framework uses cognition power for selecting the most relevant objects. A real case study was also presented to validate the framework more efficiently [35].
- CIoT was adopted to make applications work intelligently, by learning

through the actions, performing active decision making, and finally responding to the environment with adaptive reaction [38].

- Another advancement in CIoT are Cognitive Radio Networks (CRN), which were first coined in 1999 by Joseph Mitola [24]. CRN was re-coined by Simon Haykin by re-framing it with dynamic spectrum utilization management in 2005 [10].
- The Website Preference Scale (WSPS) was developed by learning market strategies, using CIoT technology. The idea behind WSPS is to find the most preferred content elements for a website. It helps the website developers achieve a higher number of visits [28].
- The cognitive M2M communication network was developed with a perspective of the protocol stack. It was implemented with the help of cognitive radio technology [1].
- Technological heterogeneity and complexity of IoT architecture were reduced by managing the IoT objects with cognitive management framework. The framework targeted improvement in energy-efficiency by selecting the most suitable objects from the set of diversified ones [8].
- The real objects and virtual objects are used for providing IoT services. These objects are packed and framed with constraints to predict future potentials of a real life problem. The framework was implemented with the help of reasoning and machine learning paradigms [29]. Another proposal that uses a machine learning model and neural network to forecast the future market trends [37].

2.3 CIoT Background

Internet is the network which connects the system to the central server for exchange of information and services. It is expanding at a rapid scale. The World Wide Web (WWW) gives a better meaning to the Internet and transformed the information in the form of websites. The web was developed by a scientist named Tim Berners-Lee. Internetlivestats, an Internet service company, conducts surveys periodically and as of today, the Internet users count reached 4,096,718,862, the count of websites is 1,938,897,174, and Google searches in a day reach 5,787,475,169. This survey proves how the world depends on the Internet. The Internet is now transforming from a network to the things presented in Figure 2.1 and discussed below.

1. Internet of Network (IoN): First phase of the Internet, i.e., IoN came with the development of the Advanced Research Projects Agency Network (ARPANET) in 1969. It followed packet-switching methodology, which

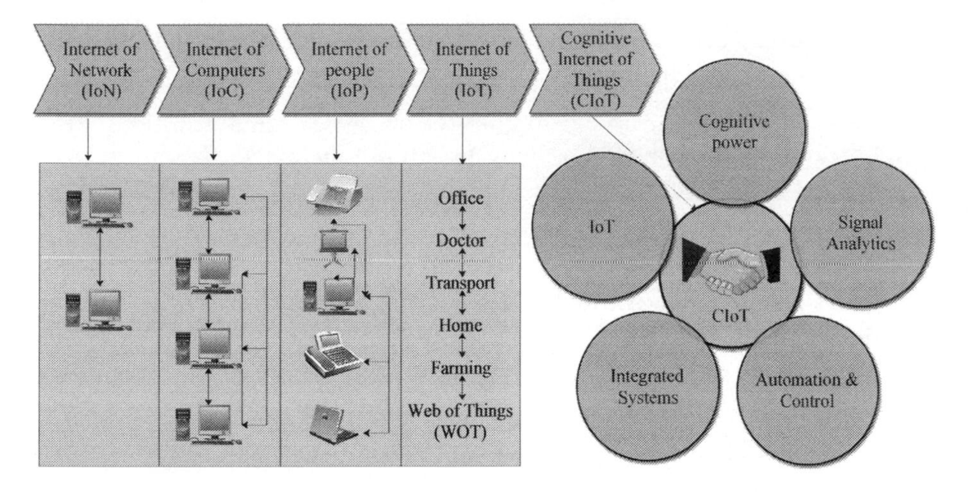

FIGURE 2.1: Evolution of Internet and its advancements

was designed by a team of scientists, namely Lawrence Roberts, Donald Davies, Paul Baran, and Leonard Kleinrock. IoN connected multiple computers over a single network.

2. Internet of Computers (IoC): The TCP/IP protocol was adopted for communication on January 1, 1983, which was developed by Robert Kahn and Vint Cerf. It gave rise to the concept of a network for transmitting the data over multiple networks.

3. Internet of People (IoP): IoP was founded by a community, known as the Fermat community. The founder of the IoP project was Luis Fernando Molina, accompanied by Lan Tschirky. It leads to the transfer of data from one device to another without using multiple servers and leads to privacy and security. It was the outperforming network for mobile phones, which lead to social connectivity amongst people.

4. Internet of Things (IoT): IoT emerged through communication among multiple objects with minimum human intervention. The discussion regarding the concepts of connectivity among smart devices started in 1982 and was introduced by Kevin Ashton in 1999. IoT is now applicable in various fields such as transportation, agriculture, education, medical, and industry.

5. Cognitive Internet of Things (CIoT): Cognitive Internet of Things (CIoT) is another paradigm of IoT developed to enhance the capabilities of intelligence in IoT objects where these objects can take independent decisions in any environment. The technologies like machine learning, natural language processing, etc. are used for processing and decision making by the automated CIoT systems.

2.4 CIoT Elements

A human consists of different body parts that enable survival in this world, for example, ears to listen, eyes to see, tongue to taste, mouth to speak, and skin to feel. A computer is a combination of various input/output devices to perform basic tasks assigned by a human. Let us take a scenario, a student is preparing for an examination. The necessary elements required to perform these tasks are eyes to read, translator to read different types of inputs, and a brain to store and process the input. Similarly, a cognition system is configured in such a manner that it can listen to the speech of a human with the help of a speech recognition system, which is interpreted by the technology known as natural language processing (NLP). Different elements are required to build the successful CIoT system which are categorized into three major fields as discussed below.

2.4.1 Sensors

The data to be processed is either entered by a human into the system or is captured by the system. The data gathering through an input device is an uncomplicated task and can be easily performed, whereas the data which is captured with the help of *sensors, RFID, and Bluetooth* is a bit of a tedious job to retrieve the results. Industrialists and professionals have CIoT system installed at their workplace or remote places. So, it is the responsibility of the system to maintain data integrity and security constraints. Nowadays, biometric systems are installed at various working areas, such as in schools for attendance, in industries for recognition of employees, and so on. The speech recognition system is also a major contribution to the CIoT data. Starting from the siri system by Apple and upto the deep learning based biometric systems, speech recognition software is considered valuable and reliable.

2.4.2 Machine Learning

The input/data/IoT data enters the CIoT system with the help of sensors. Machine learning is used for processing the input data to produce the output data. It is the field of AI, in which a system learns from the past data and the experience gained is applied in solving the problem. It is the learning methodology which learns from the data to process other data. It is the method by which the CIoT system gets periodically updated and learns from experience.

There are two basic categories of machine learning models, i.e., supervised learning and unsupervised learning. In supervised learning, the models are trained with labels, and in unsupervised learning, the models are trained without labels. Neural networks, decision trees, support vector machines, Bayesian, bagged multivariate adaptive regression splines (MARS), bagged carts, and boosted linear models are a few machine learning models. Prediction by a machine learning model follows the defined approach. Machine learning is used in industry experiments to predict real-time data. Weather forecasting is one of the IoT applications where machine learning is applied.

2.4.3 Cloud Storage

The cognitive system requires many elements to gather information. The various sources for gathering information are such as sensor data, structured or unstructured data and many more. Cloud technology is not a new technology in the eye of scientists. Time-sharing system tore off with the mainframe systems, which was expensive and difficult to install. As the CIoT system provides the number of atomic services, cloud technology has been adopted. This modern computing system reduces capital cost and maintenance too. A variety of cloud solutions have emerged over the last decade. Public clouds and private clouds both provide the same level of security, but with different infrastructure. This cloud storage offers capabilities for storing data as well as backup and restore functions.

2.5 How Do Cognitive Devices Act as Human Assistants?

With the invention of the Internet, humans now depend on it to perform various tasks. Text books and notes have now become only a part of school. IoT devices have drastically changed human life. All of us, from children to the elderly, uses IoT devices as a part of their life. This is because of the development and progress of innovations made by researchers. Same is the case of CIoT devices, a cognitive toothbrush, a cognitive car, a cognitive home device, cognitive equipment, a cognitive helmet, are in common use. Now, CIoT devices are working as human assistants in below-discussed scenarios.

1. Health Assistant: The complete success of life for humans depends upon the health constraint. According to India times in the time span of 2006 to 2050, the population of india will grow by 40 percent, in which the senior citizen population is growing rapidly with 270 percent. The report of economic time states that the senior citizen population will rise to 32 crores by 2050. You can observe the activities of senior citizens in your family. With age, they forget to do small tasks, like taking medicine on time and turning off the electrical equipment. But, the question is, how CIoT can resolve these issues? Senior citizens can take advantage of CIoT devices for delivering services to them. They can use a device which keeps track of daily activities, such as a fit band or heart tracker to track physical

activities. The house can be equipped with sensors to keep track of every single activity performed, such as whether electric equipment is in use or not, the door is closed or opened, if food was eaten or not, and whether medicine was taken on time or not. This is the best way to keep track of daily activities.

2. Shopping Assistant: Shopping is a necessity of survival. By buying the vegetables, one can prepare the food. Similarly, clothes, medicines, and household things are all required. I will give you an example that happened to me recently. I went to the store to buy the diapers for my child. The cost demanded by the shopkeeper was 910 Indian rupees after deducting 100 rupees as 10% discount for 72 diapers. Now, you can calculate the cost of a single diaper. I browsed online and checked the price of diapers elsewhere. I could get them for 975 rupees for 120 diapers after deducting 623 rupees with a 39% discount. This was possible for me because of online shopping sites. There are many shopping websites by which one can compare prices and offers. It would take 30 days to visit 100 shops and remember these many offers. Whereas 100 websites can be compared in 30 minutes. The online assistant has removed the hurdle of such an exhausting act. Vehicle insurance websites developers are so kind, that our single click on their websites lead to hundreds of phone calls from various insurance assistants. This is how CIoT devices can work as a shopping assistant.

3. Personal, Social, and Education Assistant: What if you forget your friend's birthday or forget to attend the event you already have registered for. It does not lead to any loss, but of course remembering such activities would make you more socially active. I hope you all have guessed what it means. Yes, Facebook is social networking which makes us socially and personally active. All thanks to *Mark Zuckerberg*, it reminds you about the events you wish to attend, your activities after some time (1 month, 6 months, 1 year) and works as a reminder. There are many other social networking websites as well.

   Education websites are a great help for students and learners. The knowledge of a student is not limited to a chapter of a book. Google, an evergreen search engine, provides a large volume of information gathered across the world. Research material used by the researchers is also accessible on the Internet. Research papers, experimental data, deep information about related topics can be easily found. Websites such as Google Scholar can help researchers to get updates related to recent publications.

2.6 Machine-to-machine Interfaces

Consider the scenario of traffic jam, where all the vehicles are cognitive devices and are trying to get out of it as soon as possible. Though the repetition is one of the qualities of cognitive devices they may not end up with just blowing the horns at each other. Imagine the home fully equipped with cognitive devices which are not synchronized. All these devices are programmed with turning off the electrical equipment at the same time can perform the same task together. So, there is a need of network that communicates with each other to act smartly. Machines perform differently from humans. They also communicate with each other with some defined rules. This is known as machine-to-machine conversation (M2M). There are various aspects of machine-to-machine communication.

2.6.1 Language

It is important to communicate with each other to perform various activities. Like a teacher communicating with a student using board and chalk, professionals use email for communication, or a child may touch their mother for communication. Similarly, machines communicate with each other using some language. The language used by the machines has some mechanism, popularly called a model. A famous model that serves the purpose of M2M is Open System Interconnect (OSI). There are seven layers in this model.

1. Application: This layer is responsible for establishing communications among users. It helps to provide the basic services for communication such as file transfer and e-mail. The successful protocols working at this layer include Simple Mail Transfer Protocol (SMTP), HyperText Transfer Protocol (HTTP), and File Transfer Protocol (FTP).

2. Presentation: This layer acts as a middle agent between a network and an application. Character encoding, data compression, and encryption/decryption are a few of the services provided by this layer. For example, this layer can translate the coding that represents the data when communicating with a remote system made by a different vendor.

3. Session: The multiple sessions between the applications are controlled by this layer. The role of this layer is mainly observed in case of wireless networks in which the session gets expired while encountering interference. The suspension of such networks is initiated by this point.

4. Transport: It provides the mechanism which is responsible for maintenance, establishment, and termination of circuits. These circuits are used to make the connections between network applications from one end of the communications circuit to another (such as between the web browser on

a laptop to a web page on a server). Protocols such as the Transmission Control Protocol (TCP) operate at this layer.

5. Network: This layer is responsible for logical addressing to predict the path. This layer transmits the packets from source to destination. The packet delivery is ensured by validating the destination. The protocol working in this layer is the Internet Protocol (IP), which controls the network.

6. Data Link: This layer is responsible for the transmission of data frames between the nodes. These nodes are connected with the physical layer. The error control and synchronization are the mechanisms controlled at this layer. The transmission of frames is controlled in case of wireless networks, by controlling the air and errors while transmitting.

7. Physical: This is where the actual transmission takes place. The physical transfer materials used at this layer include coaxial cables, radio waves, optical fibers, and infrared light.

2.6.2 Interpersonal Relationship

Machines can also work in collaboration with each other like humans. But before going into the process of communication, they must identify each other in their respective roles and thus can conclude the relationships among them. Sometimes, the machines are physically connected so that they need not search much for others' identification. Some type of security constraint is used by the devices that may be in encrypted form and is in the form of username and password. There can be thousands of devices; it is very difficult to remember and store the credentials. Three types of relationships are maintained by cognitive devices as discussed below:

1. Trusted relation: When the cognitive devices belong to the same organization (home, office, network), they may form a trusted peer. This kind of relation when built leads to an easy flow of data. The devices need not verify each other's identity before communication.

2. No relation: When the cognitive devices are not connected by any network but are socially active, they can communicate with each other by following the rules and regulations. For instance, while crossing though traffic lights, cars can give indications to each other using headlights.

3. Inverse relation: Cognitive devices also tend to behave differently with different devices. If the cognitive devices can communicate in favor of each other, then they can communicate to harm each other as well. It can be detected by fraudulent activities of data exchange.

2.7 Man-to-machine Communication

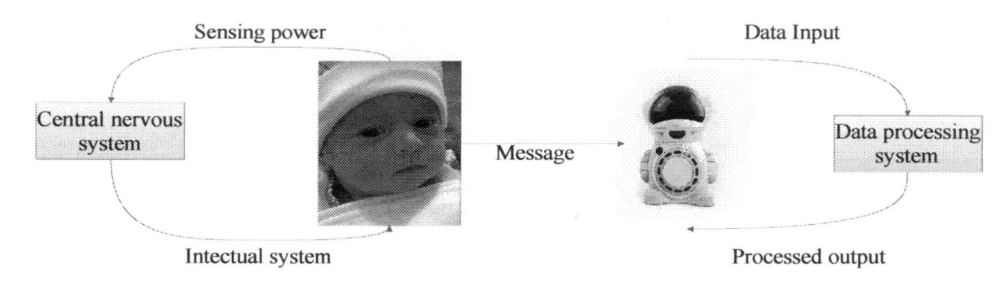

FIGURE 2.2: Communication between human and machine

This concept gives me plenty of information to implement in real life. When my child was a few months old, I started purchasing toys like rattles, and every single day my child seem to be learning. After the span of one month, the child starts and learns new activities and expects a new toy. For the next six months, I just observed the requirements and wished there could be a toy which will grow with my son and satisfy his needs. A wonderful thing happened when I found such a toy. All thanks to Cognitoys. These are touch-enabled toys which provide the feel to a human while communicating with the kid. These toys are meant to act smartly when the child changes their feelings. At every age of the child, this toy senses the growth of the child and then reacts accordingly. The toy tends to play with the child, educate the child, and guide the child in the right direction. The best feature of this toy is the ability to store and extract the information from the data provided by the parents while installing it. The continuous acquisition is the quality maintained by the toy. Although the toy is connected with the Internet for reacting it also make the use of CIoT technologies for making decisions. Figure 2.2 presented Cognitoys communication with the kid.

2.8 Machine-to-web Communication(M2W)

Previously, people try to solve their physical problems by meditating or by trying home remedies. In today's era, the physical problems of human are handled by machines. These machines/devices are developed in a dedicated way to solve specific functions. For instance, the coffee maker reduces human efforts by providing an automatic brewing system. Some devices are smart and intelligent, so can be called cognitive devices and result in a resourceful response. The question arises that a machine gets the solution to an unknown problem. It is all due to the largest repository known as the World Wide Web

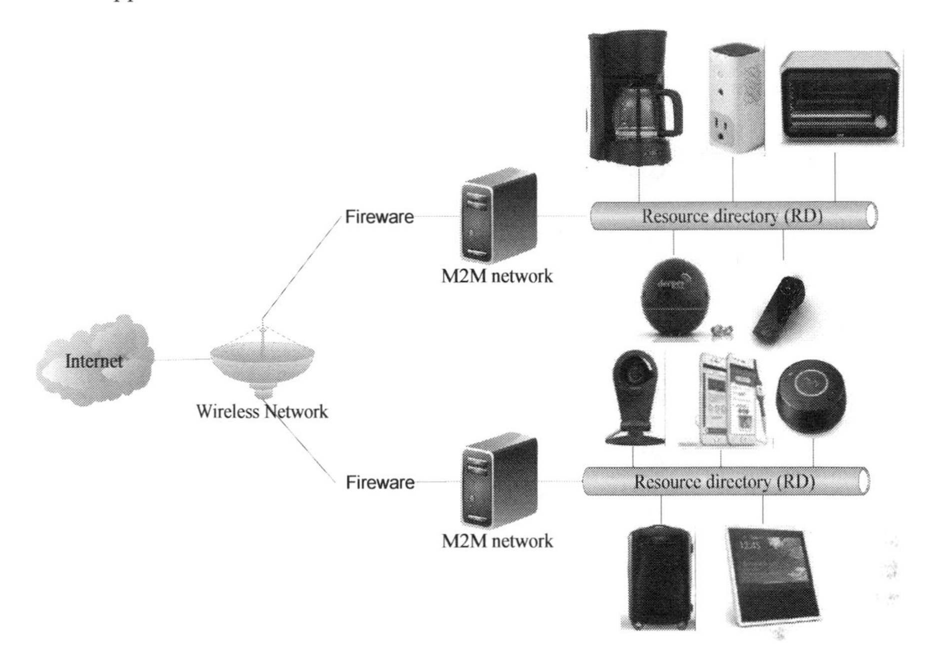

FIGURE 2.3: Process of machine-to-web communication

(WWW). The equipment can retrieve data from the repository with the ability of the Internet. However, the search engine is the platform for answering all the user's query. The web crawler is used to search the web for the appropriate answers of the query. It is not efficient for the devices to stay connected to the Internet all the time, as it consumes battery, costs, and reduces security. The resource directory (RD) comes into action. The devices firstly search the RD which in return transfers the query in the network to search engine. The mechanism followed by the web devices is presented in Figure 2.3.

2.9 CIoT Applications

With the immense development of computer systems and softwares, the human life is getting relaxed from the paper work. AI and IoT plays an important role in this advanced cognitive environment. The machines have now become more efficient and adaptive. These machines can now augment humans at every phase of life. This ability is due to the collaboration of cognitive mechanisms with AI and IoT technology. The harmony thus achieved has made some routine activities just a game of touching a button. The technology of AI and IoT will be replaced with cognitive AI and cognitive IoT shortly. A kid, a teenager, an aged man, all are using the devices in one or the other

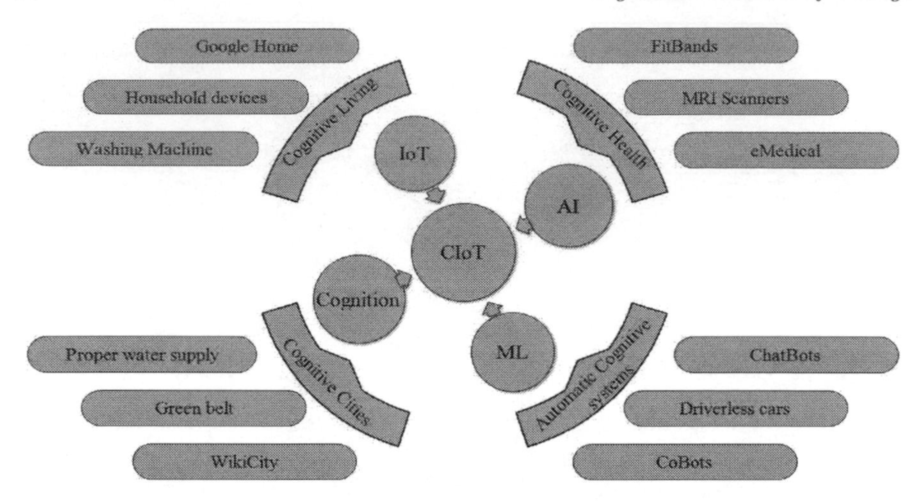

FIGURE 2.4: CIoT applications

way. Just to brief you with an example, when my child opened his eyes, it was in the front of advance technologies. First was the camera to capture that memorable moment, second was medical equipment to sense the baby's body temperature, weight, eyes, and height; third was the record book. What was expected upon the baby's arrival? Was the physical examination of the baby mandatory? All the devices were equipped to play an important role. The necessity of these technical devices was just to analyze the cognitive development of the new arrival. With this example, we can easily realize the importance of cognitive systems. These systems significantly play a major role in the field of healthcare, industry, and education as presented in Figure 2.4. The applicable fields of cognitive systems are discussed below.

2.9.1 Cognitive Living

I still remember the time when we used to make so many efforts to get morning mean. But now, bed tea has become the start of the day beverage. There is a drastic change in the lifestyle of every living organism in this world. Humans, animals, even plants are using cognitive systems in one or the other way. From the early morning until midnight, we are surrounded by smart devices. Smart homes, like Google Home, are one such a cognitive device which have changed every activity in a much advanced and technical manner. Let us understand the activities happening in the home surrounded by cognitive devices. The complete home devices can be installed with the help of a single device known as Google Home. Whenever the person enters the home, the electricity main power turns on. While, entering into a room, the temperature of the room gets moderated by the smart air conditioner. Every device senses the presence of the person with the support of sensors and actuators. This is how cognitive devices have made the life of humans easier and smarter. For a healthy person, it may not play much important role, but for an elderly, cognitive device plays an important role from every aspect, as it reduces the dependencies. Household devices are now developed and installed in such a manner that they remain connected to the Internet. The recognition power of cognitive devices work by fingertips, face, voice or sometimes even with footsteps. These devices can also be controlled via mobile device from a remote location. These devices are working smartly, but how they are working in cognition fashion? Let's take an example of a basic device which we use in daily life, a washing machine. This machine washes the clothes, rinses them, and then dries them. This is smartness, but the activities are done by the machine using intelligence power such as identifying cloth pattern, and then deciding the washing pattern is the cognition ability. This kind of washing pattern decision can reduce water wastage and electricity supply.

2.9.2 Cognitive Cities

Back to the ages, when we used to collect water from remote places to perform day-to-day activities. Now, we can enjoy an unlimited supply of water wherever we are. How is possible? It is not a blessing but the work of engineers. They have made it possible to control various parameters of living in terms of infrastructure. Let's go through the Smart Cities Mission by the Ministry of Housing and Urban Affairs, Government of India. I consider this scenario as it is meant to improve the Indian cities with smartness, thus it can be called a cognitive city. The motivation of this project includes supplying adequate water, supplying adequate electricity, managing waste with proper degradation, housing facilities to accommodate the maximum of house hold devices, maintaining efficient interconnectivity of these devices with Internet. These are the facilities and concepts that make cities smarter. It has been estimated that 70% of the world population will live in an urban area by 2050. Cognitive cities are smarter and provide advanced facilities like computerized car parking. As the cognition system has the power of remembrance, it can easily capture traffic data to predict the traffic load. So, the concept derived from this is the WikiCity, just like Wikipedia. Wikipedia contains information from all around the world, but the WikiCity contains information about the city. The temperature, traffic load, green belt, cycling area, and pollution are the different parameters recorded. It helps to retrieve the information about the particular city from any region. The WikiCity has been implemented in countries like Spain, and South Korea.

2.9.3 Cognitive Health

Health is important for a happy life. As the world is getting advanced in producing new food products, simultaneously it is leading to harmful diseases as well. The detection of these diseases is a typical research area for the doctors.

Advanced medical technologies have been developed to such an extent than many dangerous diseases are curable now, such as some forms of cancer. Single pieces of medical equipment can monitor the heartbeat, oxygen level in the body, count the blood cells in the body, measure the body mass and temperature, and are typically known as scanners. Some of the medical facilities can be availed of remotely also, typically known as remote medical centers. The advancement in the technologies lets you order medicines by consulting online and receive medicines at your doorstep. Many other devices such as fit bands helps to monitor daily activities and keep track of your health.

2.9.4 Auto-casting and Auto-reacting Cognition Systems

There exist many cognitive systems which sense the environment by itself and accordingly respond with some action. Such systems have a built-in conversion system which converts the input into the machine-desirable form and processes it for the final output. This mechanism is handled and performed by the system itself. For instance, the car senses the state of the driver and takes appropriate action when needed. The cognitive coffee maker senses the mood of the person and prepares the coffee accordingly, either strong, light, or normal. This is how cognitive systems work. Here is the list of a few cognitive systems which we use repeatedly.

1. CoBots: The CoBot is the cognitive device developed by the mechanical engineer J. Edward Colgate in 1996 and patented by Michael Peshkin in 1999. They work in collaboration with humans for assistance and are easy to program and install. They became the second generation for the manufacturing industry.

2. Driver-less cognitive car: This seems to be a dream, but it is reality. The car can run without a driver. This has been possible with the help of building tools of IoT, AI, and cognition technology. This cognitive car can sense the surroundings, the passengers, and the route. It follows the path which is directed by GPS by fixing the location at the positioning time (the time passengers sit in the car and give command). The traffic light sensors are programmed in such a way that the car can differentiate between red and color. This cognitive car can communicate with other cognitive cars with the help of messages.

3. Chatbots: "The number you are trying to reach is currently busy, please try again later." This commonly heard message is a Chatbot, which is programmed in such a manner that whenever the person is already communicating over a telephonic network, this message is sent by the subscriber. Such types of recorded message, mimicking the human voice, are known as Chatbots. For instance, Apple introduced Siri which talks like a friend and replies in a similar manner. Even the customer care call services, engage the customers with current offers and schemes, until the service executive is available to talk.

2.10 Summary and What's Next?

CIoT, the emerging technology from AI and IoT, and CIoT devices and technology, are discussed in this chapter. We discussed the history of CIoT focusing on how the technology has driven the software enhanced from the Internet to IoT. Covering all the CIoT elements, we emphasised machine learning. CIoT helps humans by working as an assistant in various fields such as health, shopping, personal life, education, and social life. All aspects of machine-to-machine communication are elaborated on, which covers language and interpersonal relationships. Man-to-machine communication lead to the development of Cognitoys, which drastically changed the world for kids. Covering all the CIoT applications, along with the auto-casting and auto-reacting cognition systems, here we summarize the chapter.

In this chapter, the data collected from the CIoT system has been explained. In the next chapter, the mining process of data is elaborated on in which data is collected from the CIoT devices.

2.11 Exercises

Questions

1. What are cognitive devices? How are different from IoT devices?
2. What do you mean by CIoT? Elaborate on it with some applications.
3. How did CIoT evolve?
4. What are the CIoT elements?
5. What is the role of machine learning in CIoT?
6. How i the CIoT data stored?
7. What are the different parameters required in M2M communication?
8. How does man communicate with machine?
9. What is meant by M2W network? Explain it in detail.
10. Explain auto-casting and auto-reacting systems.

Multiple Choice Questions

1: What are CoBots?

(a) Cognitive devices
(b) IoT devices
(c) Kind of robot
(d) None of these

2: Which type of technologies are built in the development of cognition cars?

(a) IoT
(b) AI with cognition power
(c) GPS
(d) All of them

3: What are Chatbots?

(a) Cognition radios
(b) Cognition recorders
(c) Cognition sensors
(d) None of the above

4: Among these, which technology has been considered as the secret of happy living?

(a) Cognitive robots
(b) Cognitive health
(c) Chatbots
(d) None of the above

5: Which website provides us with information regarding cognitive cities?

(a) Wikipedia
(b) WikiCity
(c) Google
(d) None of them

6: Which kind of device is the Google Home?

(a) Cognitive device
(b) IoT device
(c) Auto-casting device
(d) All of the above

7: Which system of Cognitoys helps to respond to a child?

(a) Data processing system **(b)** Central nervous system

(c) Cogni touching power **(d)** None of the above

8: Which type of cognitive relation among cognitive devices can harm each other?

(a) Trusted relation **(b)** Distrusted relation

(c) Inverse relation **(d)** None of them

9: Which type of model is maintained by cognitive devices in M2M communication?

(a) OSI **(b)** TCP/IP

(c) Cognimode **(d)** None of above

10: Why is machine learning adopted in cognitive devices?

(a) For predicting the unseen **(b)** For improving the performance

(c) For development **(d)** For applications

11: Which technology is adopted for the development of Internet?

(a) Network **(b)** Satellite

(c) Wired communication system **(d)** Bluetooth

12: Which system has the highest storage capacity?

(a) Time sharing system **(b)** Mainframe system

(c) Embedded system **(d)** CIoT system

13: An IoT device with the cognition power is known as a:

(a) Intelligent device
(b) Cognitive device
(c) Smart device
(d) All of the above

14: How does a human remain socially active in today's era?

(a) Cognitive device
(b) IoT device
(c) Social networking website
(d) All of the above

15: What kind of relation is maintained by cars at traffic lights?

(a) Trusted relation
(b) No relation
(c) Inverse relation
(d) None of the above

16: From where does the machine search the information for answering the user's query?

(a) World Wide Web
(b) Resource Directory (RD)
(c) Temporary buffer
(d) None of the above

17: The Siri system is a:

(a) Cognitive device
(b) IoT device
(c) Chatbot
(d) CoBot

18: SMTP is the protocol embedded in the OSI model at:

(a) Presentation layer
(b) Transport
(c) Application
(d) Network

19: Cognitoy is the example depicting the communication between:

(a) Machine to machine

(b) Man to machine

(c) Machine to web

(d) None of the above

20: A scanner is a:

(a) Cognitive living device

(b) Cognitive health device

(c) Auto-casting cognition system

(d) None of the above

3

Data Mining in IoT

It seems everything is connected in this world with the Internet. This is due to the growth of the Internet of Things (IoT), which has made the data available to everyone, everywhere, and anytime. We can conclude that the world is completely dependent upon the information. Information is the processed form of data. Terabytes of data are moving around us, either in the networks or the systems or the World Wide Web (WWW) or the cloud. The data is produced due to the development of computerized culture. Every industry, i.e., sales, production, the stock market, printing, software development, construction, medical, health, is using systematic data. For example, a CCTV camera is recording the footage of a complete day, which means it is storing the data in one form or the other. Similarly, researchers and scientists generate a high ratio of data in a sequence by experimenting, sensing, observing, and measuring. Millions of websites are developed instantly which are supported by the network of the WWW. If we start counting the sources of data for the websites, it would take ages to collect. Web communities have driven the world to be connected socially. The need for fetching valuable information from this tremendous amount of data has given the meaning to the concept known as data mining.

3.1 Search Engines as a Medium

The platform of Google has made the process of information retrieval easy and convenient. I think the search engine is not a new term for all of us. Let me elaborate, a search engine is the platform for retrieving the data stored at WWW by just firing a query. The query entered by the user is exactly what is required. Based on the query, the search engine returns the web pages that contain the correlated information. For example, the user asks Google for information regarding webspam. The user enters 'webspam' in the Google search web spam. The results contain all the web pages which define the meaning of webspam as well as the web pages which contain surveys related to webspam. This example illustrates how the relevant information is extracted from the huge volume of data.

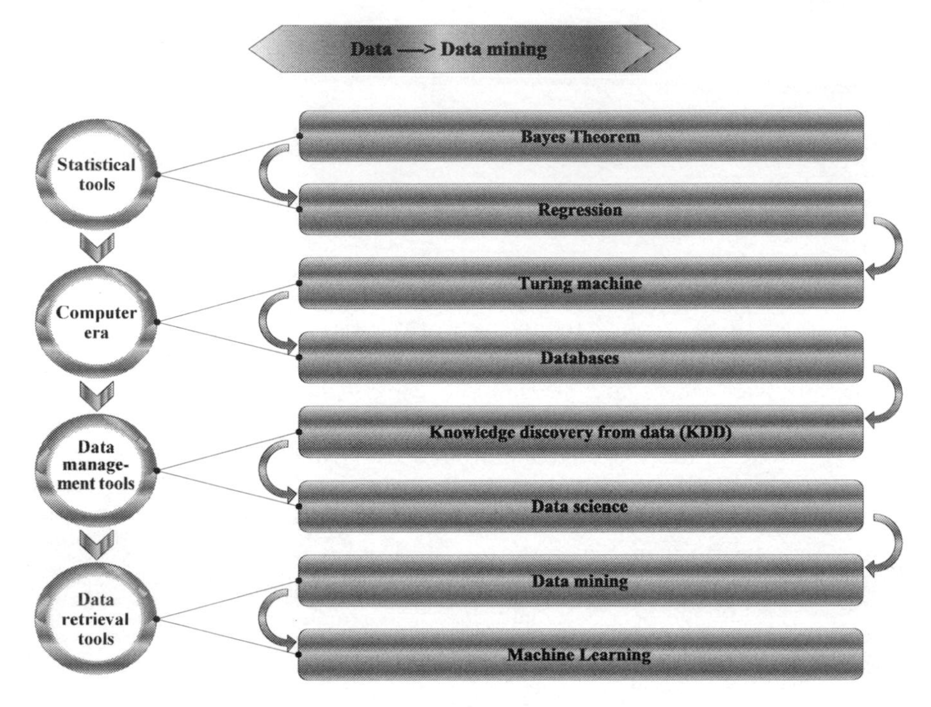

FIGURE 3.1: Stages of data mining cycle

3.2 Data Creation and Retrieval Scheme

What is data? How it is created? What are the tools required for its retrieval? How do we extract the information from the data? These issues can be understood by looking back into the history of data. The cycle of data to the database and then how data mining came into the action is presented in Figure 3.1 and elaborated below:

- Bayes Theorem: During the time, when the data was just estimated, gave *Thomas Bayes* a new idea in 1763. He came up with Bayes theorem. Bayes uses the previous information to guess the new information. What is the probability of re-occurrence of the event or how can we draw the new information from the existing information? By which, Bayes concluded that previous information with a new belief produces the new information. Bayes never thought to publish his work; it was done by one of his friends *Richard Price*. But the theorem became popular when *Pierre Simon Laplace* (scientist) made this theorem applicable in the scientific problem. This method got attention from statisticians also. This is how the data was given meaning and stored in files.

- Regression: A new analysis of data was made by *Legendre* in 1805 with the concept of regression. It was refined by Karl Gauss in 1809. The first form of regression was the method of least squares. It was further enhanced by Gauss in 1821 with the invention of a new theorem named the Gauss-Markov theorem. The term 'regression' was invented by Francis Galton in the nineteenth century to typify a biological phenomenon.

- Turing Machine: The computation machine, which is popularly called a model of computation, is also known as the Turing machine. It was invented by Alan Turing in 1936. This machine was developed to reduce human efforts. It has a read-write head and tape which is used for primary storage. The six operations which can be performed by Turing machine are left move, right move, read, write, halt, and change state. It seems to be a simple machine, but it can perform millions of operations without any hurdle. This Turing machine led to the development of modern electronic computers. This is how the file system storage moved to database storage systems.

- Databases: Charles W. Bachman invented the system of a database. It is the system to store, add, delete, or modify the data items from and in one place. As the use of computers became common in the mid-1960s, the demand for a systematic database was also raised. It gave rise to the development of a new database system known as Common Business Oriented Language (COBOL). COBOL became famous in 1971 and was popularly called a "CODASYL approach." Eugene Wong and Michael Stonebraker developed a query language known as QUEL in 1973, under the project INGRES (Interactive Graphics and Retrieval System). In 1974, SQL was developed with the existing technology called QUEL. This chain continued, and the data is easily stored by industries in the form of rows and tuples.

- Knowledge Discovery in Databases (KDD): The research started for optimizing the results retrieved from the database. In history, it was found that it could be done with the help of machine learning. Gregory Piatetsky-Shapiro proposed KDD, while working in GTE Laboratories on CALIDA. After attending a few conferences, he realized data discovery is a good idea for solving the problems of data retrieval in databases. Knowledge discovery is somehow similar to data mining. Knowledge discovery deals with database creation and data retrieval both, whereas data mining refers to data abstraction. The various steps involved in KDD are shown in Figure 3.2.

- Data Science: Since 2000, data was created using databases and extracted using KDD. Data science comes into its role. Data science is the scientific method used to extract the unstructured and structured data from the database. In 2001, 'data science', the term, was coined by William S. Cleveland. Data science is the multidisciplinary task which includes data

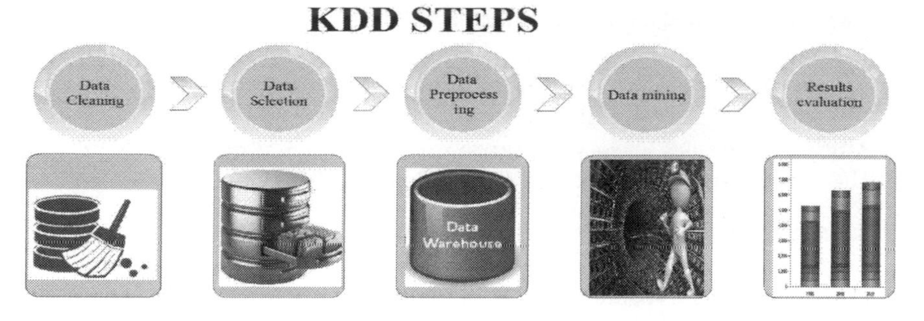

FIGURE 3.2: Knowledge discovery from data steps

visualization, data analytics, data sciences, statistics analysis, natural language processing, and computational sciences.

- Data Mining: The analysis of classical statistics is known as data mining in artificial intelligence. However, the two terms data science and data mining are used interchangeably at the Internet, but there is alot of difference among the two. Please refer to Table 3.1 for better understanding.

TABLE 3.1: Differences between data mining and data science

Difference	**Data mining**	**Data science**
Scope	It is a technique	It is a complete field
Process	It is the data retreival process	It is the scientific process
Input	It processes the structured data	It processes any form of data
Outcome	It finds the hidden trends in data for better understanding of data	It is the process of statistical analysis which involves the scientific experiments
Purpose	It is one of the steps in KDD	It is the complete process from data collection to data analysis

3.3 Data Mining

How can we mine the information from a large amount of data which is produced by IoT devices? Or how can we mine the knowledge from the data as shown in Figure 3.3? We have studied how the data is collected in IoT using sensors. It is quite obvious that using popular data handling schemes such as cloud computing, a large number of knowledgeable patterns are detected.

FIGURE 3.3: Knowledge mining from data

Data extraction is a module that needs special attention. However, the different parameters considered for handling the data are summarized in Figure 3.4 and are described as follows.

1. Data devices: There are numerous IoT devices, which can sense the environment with sensors and produce the data in one form or the other. These devices generate the raw data, which needs to be processed.

2. Data gathering: The data produced by IoT devices are gathered at a particular place by following the procedure of parsing, analyzing, collecting, and merging. This process is responsible for efficient data extraction.

3. Data processing: Data processing is the process of converting raw data into information. Nowadays, open-source solutions are used for this purpose, such as Hadoop.

4. Data services: Data services, such as data storage, data presentation, data extraction, and data binding, are the different data services used.

5. Data tools: Data tools are responsible for data integration, data mining, and data sharing. It starts with the data at network and grows with the advanced tool such as machine learning.

3.3.1 Data Mining Functions

Suppose you are asked to find the most talented student in your class. What you will do? First of all, you will find the various data sources such as student

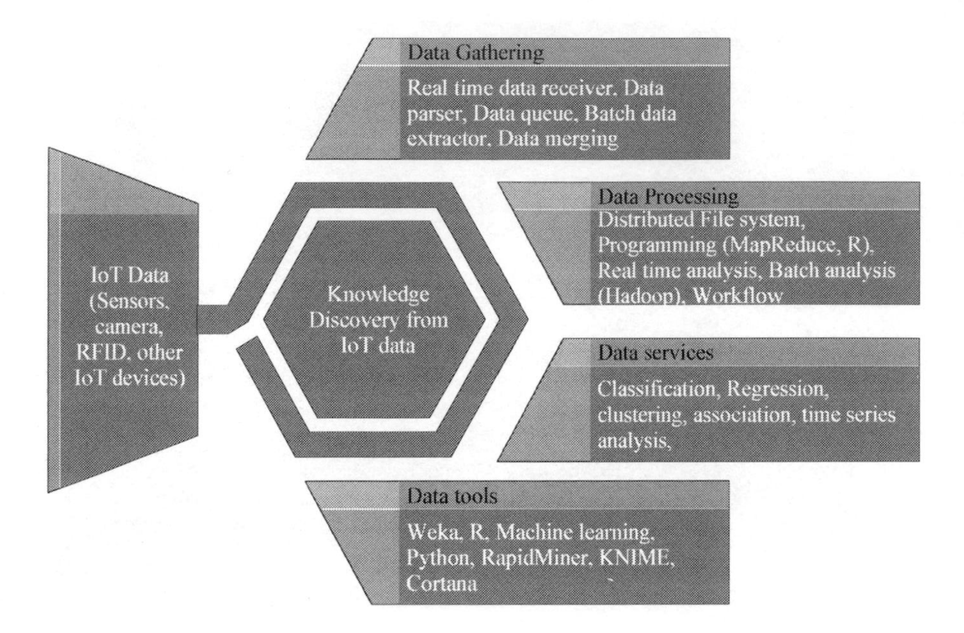

FIGURE 3.4: Representation of data mining in IoT

record book. After collecting the data, you can process it for the essential information. Once the data is processed, it is stored, and then various data services can be performed with different data tools. The process of data mining ends with the outcome, which highly depends upon its function. The different data mining functions are as follows.

3.3.1.1 Classification

It is the process of identification of data classes that can be used for the prediction of data class with unknown labels. The procedure is defined in such a manner that this process helps target class to predict the data for each case. Different methods exist for the classification of data such as instance-based, rule-based, and decision-based. All these methods use different approaches for the categorization of data. Based on these methods, different models can be designed, like neural networks, hierarchical classification, support vector machines, and many more. A few of these methods are discussed below:

1. Decision Tree: Decision trees are one of the examples of supervised learning, in which the data is split into different branches. Each branch is the parameter which helps to differentiate the data. The terms used to represent tree are nodes and leaves. The leaves are the final values, whereas nodes work as the condition for splitting the data.

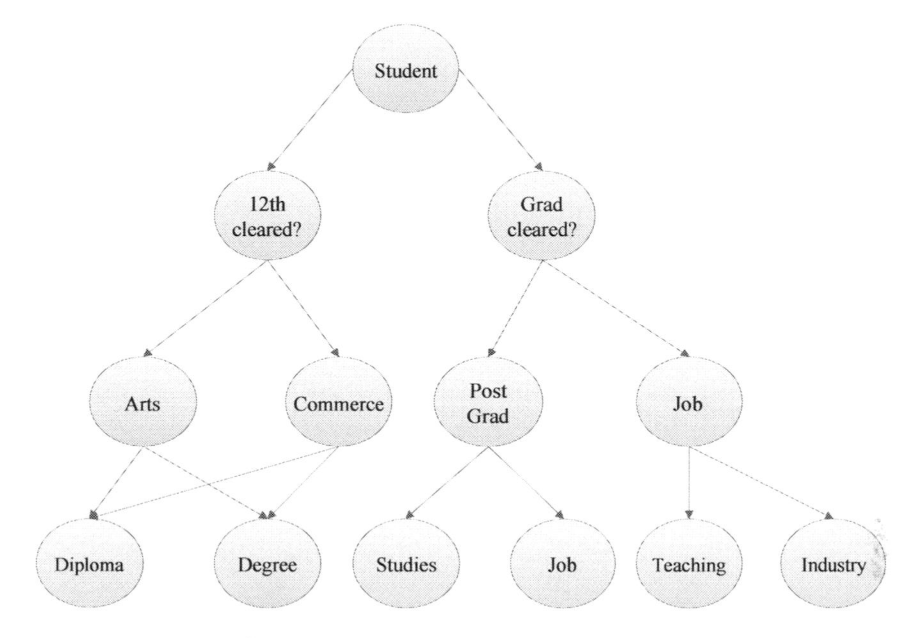

FIGURE 3.5: Tree structure representation

As shown in Figure 3.5, there are nodes like grad cleared, arts, and there are leaves as well like studies, teaching. What do you really understand by this tree? Every node has two possible leaves and with the condition satisfying in a positive or negative manner. The condition does not have any result lying in decimal precision. It states that it is a binary classification tree.

2. Neural network: A neural network is the set of multiple algorithms that are gathered together to work in a defined complicated task. The most simple architecture to understand the structure of a neural network is the perceptron.

   As shown in Figure 3.6, there are three layers, i.e., input layer, weighting layer, and activation function.

3. Bayesian networks: As decision trees are based on predefined conditions and the input to the function are fixed, there may be some unseen circumstances. In that case, the predefined conditions do not work. Here comes the role of Bayesian networks, which are based on probability. They follow the approach of conditional probability. They state that what would happen under the unseen combination of probability parameters. They give the result for unlabeled instances. The complete procedure is illustrated with an example as discussed as follows.

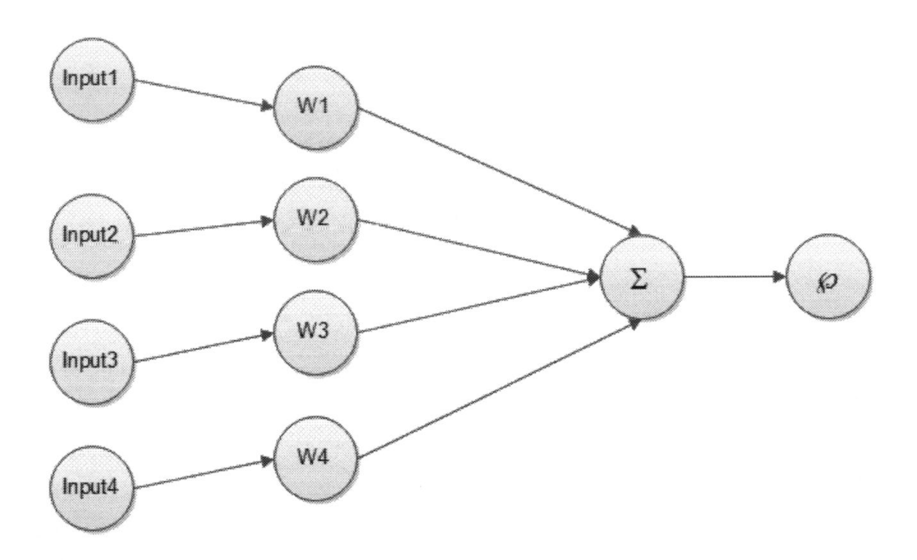

FIGURE 3.6: Perceptron architecture

Firstly, we compute the probability of the target attribute. As the target attribute, i.e., rain, can have two values only, the probability is computed as follows.

$$P(Yes) = 4/7 \tag{3.1}$$
$$P(No) = 3/7 \tag{3.2}$$

Now, we compute the probability for other attributes. Firstly, we do it for Outlook attribute.

$$P(Sunny|Yes) = 0/4 \tag{3.3}$$
$$P(Sunny|No) = 2/3 \tag{3.4}$$
$$P(Cloudy|Yes) = 1/4 \tag{3.5}$$
$$P(Cloudy|No) = 1/3 \tag{3.6}$$
$$P(Rainy|Yes) = 1/4 \tag{3.7}$$
$$P(Rainy|No) = 0/3 \tag{3.8}$$
$$P(Moderate|Yes) = 1/4 \tag{3.9}$$
$$P(Moderate|No) = 1/3 \tag{3.10}$$

Similarly we do it for atmosphere and wind attributes.

$$P(Hot|Yes) = 0/4 \quad (3.11)$$
$$P(Hot|No) = 2/3 \quad (3.12)$$
$$P(Cold|Yes) = 2/4 \quad (3.13)$$
$$P(Cold|No) = 1/3 \quad (3.14)$$
$$P(Normal|Yes) = 1/4 \quad (3.15)$$
$$P(Normal|No) = 1/3P(No|Yes) = 2/4 \quad (3.16)$$
$$P(No|No) = 3/3 \quad (3.17)$$
$$P(Yes|Yes) = 1/4 \quad (3.18)$$
$$P(Yes|No) = 1/3 \quad (3.19)$$

The probabilities from Equations 3.11 to 3.19 are computed using Table 3.2. Now, let's check for an unseen combination of parameters. Like, the day is Friday, the outlook is cloudy, the atmosphere is normal, and the wind is present. Computing the probability of rain in such circumstances using the above equations is as follows.

TABLE 3.2: Weather prediction for rain

Day	**Outlook**	**Atmosphere**	**Wind**	**Rain**
Sunday	Sunny	Hot	No	No
Monday	Cloudy	Cold	Yes	Yes
Tuesday	Rainy	Cold	No	Yes
Wednesday	Moderate	Normal	Yes	No
Thursday	Moderate	Normal	No	Yes
Friday	Sunny	Hot	No	No
Saturday	Cloudy	Cold	No	No

A = <Friday, Cloudy, Normal, Yes>
We compute the probability of both the cases that it would rain in situation or it would not rain in Equation 3.20 and Equation 3.21.

$$P(A|Yes).P(Yes) = P(Friday|Yes).P(Cloudy|Yes).P(Normal|Yes).\\ P(Yes|Yes)P(Yes) \quad (3.20)$$

$$P(A|No).P(No) = P(Friday|No).P(Cloudy|No).P(Normal|No).\\ P(Yes|No).P(No) \quad (3.21)$$

All these probabilities are already discussed; it is an exercise to calculate it.

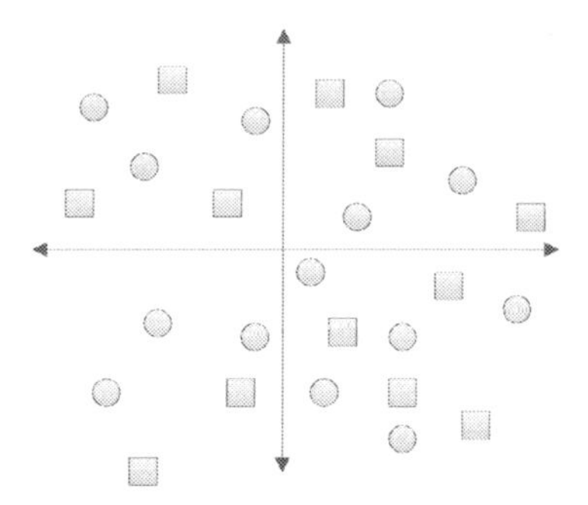

FIGURE 3.7: Data points

4. Support vector machine: A support vector machine (SVM) is officially described by a separate hyperplane as a discriminatory classifier. In other words, given the marked training information (supervised learning), an ideal hyperplane is produced by the algorithm that categorizes fresh instances. This hyperplane is a line dividing a plane into two sections in two-dimensional spaces where it lies on either side in each class.

   In summary, we can say that it separates the data into two different classes with the help of hyperplane. The data lies in all the four dimensions as shown in Figure 3.7.

   Functional margin: So, when we know about the data points, the task is to identify the different classes. This objective is achieved with the help of calculating the line function. This function is denoted as follows.

$$wx + b = 0 \tag{3.22}$$

   In Equation 3.22, the w is the weight, x is the value of coordinate, and b is the constant value. The plane not only depends upon the function but also on the coefficients. The coefficients decide the margin which is maximized for the detection of outliers.

3.3.2 Relation of Data Science with Machine Learning

It is observed that data science is developed using various tools and techniques. Machine learning plays a crucial role in their development. Development of data science applications is adopted in our daily life activities. However, machine learning was meant to make the systems learn. The system is built in such a manner that the system keeps on learning with experience. Let's take the example of the Internet search. Every search engine you use is connected to your one or the other account. Keeping an eye on your search history, it suggests frequently searched web pages automatically. The reason behind this is machine learning. Here is the list of data science applications that use machine learning as a tool.

1. Recommender systems
2. Surfing Internet
3. Image recognition
4. Speech recognition
5. Data mining applications
6. Forecasting systems
7. Natural language processing

3.4 Data Mining in IoT

It sounds like mission impossible to connect everything on earth together via Internet, but in the near future, the Internet of Things (IoT) can dramatically change our lives by making many "impossible" things possible. To many, IoT generates or captures huge data with extremely helpful and valuable information. Data mining undoubtedly can play a crucial role in making this type of scheme intelligent enough to provide services and environments that make life more convenient.

Technically, all IoT stuff can generate a data deluge containing various types of precious information. However, in recent years, technical problems and difficulties have arisen as to how to manage this data and how to dig out the helpful information. An easy taxonomy for distinguishing data types from IoT is to use "information about things" to refer to information describing stuff themselves (e.g., country, place, identity, etc.) and "information produced by things" to refer to information produced or captured by things.

Because of these features, this fresh form of information, i.e., sensor or RFID-captured information has been described as "large information." These features let us understand rapidly that big data is no longer a company hype, but something that is so real that we have to face seriously today. Organizations are expected to own more and big data in the near future from the services, applications, and platforms they provide. Besides managing these huge amounts of data, it has become a significant job for these organizations to discover information hidden in these data because this task can put them in an unassailable situation.

Since KDD has been effectively applied to different domains to discover concealed information from the data in question. It has been for years that the KDD becomes the basis of several information systems such as tracking the record for the production of lays chips. By using the following measures, KDD is naturally expected to discover "something" from IoT, and the patterns thus retrieved are discussed in Figure 3.8. The role of KDD in IoT is discussed in Figure 3.9.

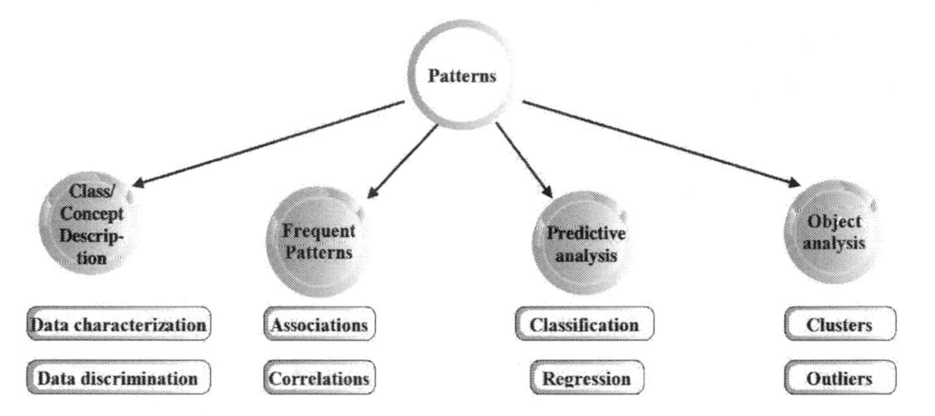

FIGURE 3.8: Patterns classification formed after data mining by KDD

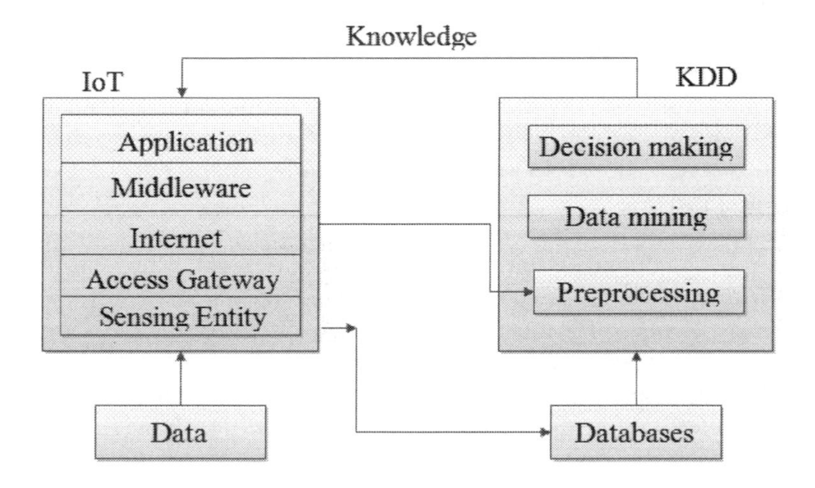

FIGURE 3.9: Role of KDD in IoT

3.5 Machine Learning in IoT

Every sub-branch of AI contributes to putting intelligence into the machines in some way or the other. IoT and machine learning both are the most upcoming fields in terms of technology. How can both be connected to produce interconnected scenarios?

1. How the data which is stored in the cloud by IoT, can be retrieved with machine learning.

2. How the data can be visualized with better insights from a large volume with the help of machine learning.

According to a survey, there are 31 billion connected IoT devices. The data transferred to the cloud increases by 27.6% per year according to IHS research. We can do various functions by using machine learning, which can generate intelligence. Thus, it makes more efficient decisions.

Machine learning technology helps us in daily routine activities. It can be visualized with the below examples:

- Shopping recommendation cart: The huge shopping complex centralizes the facilities such as billing, item verification, offer displays, and many more. The billing desktop systems are connected to a server for fetching the prices of items. However, the system records the sequences of items preferred by the customers. It saves the record which helps the following customers in purchasing the necessary items. For instance, a customer purchases items, namely, olive oil, fresh vegetables, and spices. Now, whenever any customer picks the olive oil, then the fresh vegetables and the spices, are suggested by the shopping cart. In the same manner, if fresh vegetables are chosen, then the olive oil and the spices are recommended. In this way, the system gets aware of customer choices along with preferences.

- Recipe recommendation cart: Some shopping carts using recommending systems not only store the customer preferences but also store the recipe ingredients. The purpose behind this scheme is to provide a suggestion to the customer in the course of item selection, the outcome of which leads to efficiently purchase and exploring new recipes.

Storing the current state and predicting future instances is the game of machine learning. These examples illustrate that machine learning technology is helping the advanced systems to work smartly and efficiently. In the previous chapters of IoT and CIoT, machine learning plays an important role. Let's illustrate its benefits.

- Machine learning is the technology adopted during the IoT data preprocessing by detecting anomalies in the data.

- Machine learning helps in gathering a large amount of data from industries.

- Machine learning is the tool used for predicting the IoT data.

- Machine learning performs the optimization of results, generated during data analytics in IoT. The cross-validation approach and ensemble approaches are followed for this purpose.

- Machine learning is a crucial element in CIoT data processing.
- Machine learning helps in storing the CIoT data in the cloud.

Machine learning is important for Internet technologies. Based on this fact, many researchers have attempted to implement their IoT developments in machine learning. One of the remarkable works is done by Bin et al., [4]. In this proposal, machine learning models were used for processing the IoT data. The validation of the scheme is achieved by four different data mining models, namely, the multi layer model, distributed data mining model, grid-based data mining model, and data mining model from a multi-technology integration perspective. These models were deployed at the multiple layers of IoT architecture.

Choosing the right and efficient machine learning algorithms is also an important aspect for which many researchers proposed many schemes. [22] provided IoT data analytics. The concern of this proposal is to find which machine learning model is adopted for an application. The three considerations concluded are the number of IoT devices, the number of data features, and the type of application.

IoT data is merely called smart data. Smart data is famous due to its proper meaning given by Sheth [30]. He defined the transformation of the big data into smart data, which means, "realizing productivity, efficiency, and effectiveness gains by using semantics to transform raw data into Smart Data." Once the smart data is generated, the machine learning model can be easily applied. The algorithm/model suitable for IoT applications as surveyed is listed in Table 3.3.

3.6 Summary and What's Next?

The previous two sections cover the techniques by which the data is collected by the IoT devices and CIoT devices. Now the question arises of how and what data can be retrieved. This technology, data mining is explained in this chapter. Search engines provide the platform used for accessing the information from the WWW. How the data term evolved and how it converted the world into the information era is elaborated. Machine learning plays a crucial role in the data mining of IoT. This has been explained with the application being developed by researchers. In other words, we can say machine learning in data mining is crucial.

TABLE 3.3: Machine learning models in IoT applications

Author	Year	Algorithm	Smart data	Application
kafi et al. [15]	2013	Classification	Smart traffic	Urban traffic monitoring
Qin et al. [27]	2016	Clustering	Smart traffic	Increase data abbreviation
Toshniwal et al. [34]	2013	Clustering	Smart health	Health monitoring systems
Jakkula et al. [12]	2010	Anomaly Detection	Smart Environment	Finding anomalies in IoT power data
Ni et al. [25]	2014	Support Vector Regression	Smart weather data	Forecasting the weather
Derguech et al. [6]	2014	Linear Regression	Smart market data	Predictive analytics system
Ma et al. [21]	2013	Classification and Regression Trees	Smart citizens	Passenger travel pattern detection system
Khan et al. [16]	2014	Support Vector machine	Smart market data	Classification of data streams
Do et al. [7]	2015	K-Nearest Neighbors	Smart citizens	Efficiency of the learned metric
Hu et al. [11]	2015	Naive Bayes	Smart citizens	Estimate the numbers of nodes
Shukla et al. [31]	2015	Density-Based Clustering	Smart citizens	Fraud detection system
Makkar et al. [23]	2019	Two-level Classification	Smart websites	Web spam detection system

3.7 Exercises

Questions

1. Define data mining. How do search engines help in data mining?
2. Elaborate on the evolution of data mining technology.
3. Define KDD. What are the various steps involved in KDD?
4. How is data mining different from data science?
5. How is data mining helpful in IoT?
6. What are the data tools used in the data mining of IoT?
7. What are the different knowledge patterns retrieved from data mining?
8. What is the role of machine learning in IoT?

9. Which machine learning models are applicable in IoT applications?
10. What are the different data services in the data mining of IoT?

Multiple Choice Questions

1: Why was data mining technology developed?

(a) For data retrieval **(b)** For abstracting the hidden patterns

(c) For knowledge discovery **(d)** All of the above

2: Which statistical tool started analyzing the statistics data?

(a) Bayes theorem **(b)** Turing machine

(c) Regression **(d)** All of the above

3: Who invented the Turing machine and when?

(a) Charles Babbage, 1932 **(b)** Alan Turing, 1936

(c) Charles W. Bachman, 1936 **(d)** None of the above

4: Who invented the technology of database?

(a) Charles Babbage **(b)** Alan Turing

(c) Charles W. Bachman **(d)** None of the above

5: KDD deals with:

(a) Data creation **(b)** Data retrieval

(c) Data abstraction **(d)** Data creation and data retrieval

6: Which data is known as smart data?

(a) CIoT data
(b) IoT data
(c) Processed data
(d) Data retrieved by IoT devices

7: Which algorithm was designed to generate the data of the smart environment?

(a) SVM
(b) Naive Bayes
(c) Anomaly detection
(d) Linear regression

8: The fraud detection system was developed using the algorithm:

(a) K-nearest neighbors
(b) Naive Bayes
(c) Density-based clustering
(d) Classification

9: The definition of smart data as "realizing productivity, efficiency, and effectiveness gains by using semantics to transform raw data into Smart Data," is given by:

(a) Dr. Kafi
(b) Dr. Jakkula
(c) Dr. Sheth
(d) None of above

10: Machine learning technology is used for:

(a) Detecting the anomalies in the data
(b) Predicting the IoT data
(c) Gathering the large amount of data
(d) All of the above

11: Selecting a machine learning model for developing IoT device mainly considers:

(a) Number of IoT devices

(b) Number of data features

(c) The type of application

(d) All of them

12: Which technology is used by weather forecasting systems?

(a) Machine learning

(b) Internet

(c) Artificial Intelligence

(d) None of the above

13: Which data mining function is based on probability?

(a) Classification

(b) Neural network

(c) Bayesian network

(d) Support vector machine

14: Perceptron is developed using the algorithm known as:

(a) Classification

(b) Neural network

(c) Bayesian network

(d) Support vector machine

15: The data gathering in IoT is done by:

(a) Data parsing

(b) Data analyzing

(c) Data collecting and merging

(d) All of the above

16: Which scheme is used by SQL for its development?

(a) QUEL

(b) CALIDA

(c) COBOL

(d) None of the above

17: Which scientist helped in the popularity of Bayes theorem?

(a) Thomas Bayes
(b) Richard Price
(c) Pierre Simon Laplace
(d) None of the above

18: Which is the layer in between the input layer and the output layer in a neural network?

(a) Weighting layer
(b) Activation layer
(c) Application layer
(d) None of the above

19: Decision tree formation is based upon:

(a) Data
(b) Conditions
(c) Decisions
(d) All of the above

20: Machine learning is a:

(a) Data integration tool
(b) Data mining tool
(c) Data sharing tool
(d) Data predicting tool

4

Machine Learning Techniques

Now you all are familiar with the term machine learning and its need in our life, which depends upon IoT. Machine learning is an automated learning because it automatically detects the hidden insights of the data. This growing field of computer science is mainly used in extracting the required information from a large amount of data. This is what we have learned in the previous chapter—how machine learning is used in pruning the data for performing the action in IoT devices. Leaving IoT for a second, the applications of machine learning can be easily felt by looking into the search engine result pages. The credit cards we use are also secured with machine learning tools. Smartphones are greatly embedded with machine learning technology. The smart cars we drive are also equipped with machine learning-trained tools to prevent theft and accidents. This is how machine learning revolves around us in one or the other sense. There are countless applications of machine learning. Machine learning systems not only follow the instructions coded but learn from the experience. This is what differentiates machine learning from other intelligent tools. No human being can memorize each minute value of every experiment that the machine learning can. Like face recognition, the system is trained with a machine learning algorithm that the smallest pattern should be remembered. Let's see what is learning and how the machine learning systems *learn.*

Learning: A Mechanism

Learning can be defined as an experience used for converting information into knowledge. The data thus converted is used for decision making. The machine learning algorithm considers training data as information and provides the knowledge with the acquisition of experience. The knowledge is used to perform some action, that may take different forms, depending upon the input. The process of input to transform into output depends upon the type of machine learning algorithm. The mathematical equations and science procedures are involved in this process of transformation. With every training, the system learns. This learning process is also carried in our life as well. For instance, a child when in the stage of developing taste learns new food flavors and acts accordingly in the future. If the child likes the flavor, then next time, the input is the adopted flavored food, the output is the child eats more. Another learning we are familiar with is of spam emails. Suppose you are designing an email spam detector system. You train the system with a bunch of spam emails. Now the system is trained in such a manner that next time whenever the spam email is received, it will be automatically moved to spam folder instead of the inbox folder. Now the system judges the email as spam, either with the size or keywords or font. This depends upon the methodology of a machine learning algorithm. Then, when a new e-mail arrives, the machine can check whether one of the suspicious words appears in it and predict its label accordingly. Such a system would potentially be able to predict the correct label of unseen e-mails. The machine learns or the machine learning depends upon various factors; a few of them are as listed below:

1. Category of data: The learning mechanism inherits the features of data. The learning scheme completely depends upon the data type. It is generalized to define the learning problem according to the data, like face recognition systems learn by the facial expressions. The different types of data are:

    - Vector: Sometimes, there is a need to deal with variety of instances together, which can be collectively known as vector. Spam detection systems learn the data in the form of a vector, i.e., the web page features (anchor text, body, metadata, in-links, out-links). Now the constraints in such parameters are that they may contain different scales and units. The vector parameters vary in units such as kilograms, grams, tons, units, and need to be normalized before feeding them into the system. The transformations required are thus encoded and implemented.
    - Lists: Vectors obtain different categories of parameters. Sometimes, there is a requirement of only one type of parameters. Such as web page features (link based features and content based features), in which only link based features need to be considered. So, the features like PageRank, in-link number, out-link number, are considered and together are known as a list. The same transformation operation like normalization might be required.
    - Matrices: The value corresponding to each parameter can be categorized, meaning there is a requirement of different data types known as matrices. It takes the form of row and column. This form reveals the relationship between the products. Like the name of the user and its address. Such type of data is needed when we have a similarity among the features.
    - Images: Downloading and uploading the images sometimes leads to the down-sampling of images. But it may not change its characteristics. Images might have different formats such as .jpg, .png, .giff, .tif. The parameters of the image are extracted, such as height, width, pixels, colors, bitmap, and many more. The machine learns from such parameters.
    - Videos: Videos are treated as the collection of images. The dimensions among the images are fetched in a three-dimensional array. Graphs

and trees are also used for storing this kind of information. The dimensions are treated as the vertices of a graph or a tree.

2. Type of problem: As the complete range of fields, such as engineering, science, automation, or electronics, are applying machine learning, so there are a variety of applications which may vary in size. The type of problems/applications defines the type of learning. Let us understand the difference between the problems. In the last example of a spam detection system, the problem is to find the spam web pages. Now concentrate upon the output; it takes two values that either the web page is spam or it is not. So, it is the case of binary classification. The machine adopts the binary classifier for its learning. Different types of problems are discussed below:

    - Binary classification: The problem of having a single target value, is the case of binary classification. This kind of problem is observed commonly in real life and a large set of algorithms and theories have been developed for the same. We can estimate of binary variable y from a set of pattern X. The y results in two values, which are +1 or -1. The X consists of various parameters, $x_0, x_1, x_2 ... x_n$. An example is prediction of whether a person will be able to pay a loan depending upon the credentials provided such as salary, liquid cash, royalty, and property. Another is teaching a kid whether the round object is an apple or orange. The various parameters considered for this teaching is mainly the color and the shape. These are the instances of binary classification.
    - Multi-classification: This is the extension of binary classification with the difference that the y can achieve multiple values. For instance, judging the language of the web page—whether it is French, Spanish, Telugu, or Urdu. Another instance is the spam detection system which detects whether a web page is spam, normal, non-spam, does not exist, or cannot decide. Such cases mainly occur in specific conditions.
    - Regression: This kind of learning is when the problem is supposed to produce real-valued expressions. For example, predicting the rank score of a web page while considering the parameters such as in-link, out-link, word count, anchor text, and many other web page features.

4.1 Tools to Implement Machine Learning

Now, we all are aware of the necessity of machine learning. It is the core of IoT and CIoT. In simple terms, we can say the machine learning model is a program which instructs the machine/device to produce the output based upon the input. There are many languages to execute the machine learning models; I am discussing the experiments you all can perform at your systems, not at servers. Here is the list of few languages for machine learning that you all can practice.

4.1.1 Python

This popular tool is mainly adopted by students, developers, and researchers for programming. It is a language which has in-built large number of packages and libraries. The capabilities of Python enable us to perform a specific task in a number of different ways. Matplotlib, SciPy, and Scikit-Learn are the popular libraries used by the data scientists. We can implement machine learning techniques like classification, regression, clustering, and association in Python. Python offers a ready-made framework for performing data mining tasks on large volumes of data effectively in lesser time. Optimization, to improve the efficiency of machine learning algorithms, can be done in Python. It is an open-source software available online and can be easily installed. All the installation files are available online at python.org.

4.1.2 R

R is free and open-source software used throughout this book to demonstrate the machine learning paradigms interface for data mining. R is a sophisticated statistical software package, which can be easily installed. It supports the libraries that are used for programming most of the machine learning models. R is the language which supports the basic commands of programming. R has the ability to design a complicated structure of the program with its advanced built-in functions. Once you get familiar with the basic programming in R, you can drive through the module programming required for learning machine learning. One of the great benefits of using R is that it saves the program as well as the workspace with *.Rdata.* The workspace contains all the variables, matrices, and arrays created while executing the program. We can then rerun the scripts to transform our source code automatically into information and knowledge. As we use the basic commands of R and a few machine learning concepts in R in this chapter, slowly it will become easier to learn machine learning programming. The different R functions and commands required for a program are collectively known as an R script. R language is available for GNU/Linux, Mac/OSX, and Microsoft Windows. We also note that direct interaction with R has a steeper learning curve than using GUI-based systems, but once over the hurdle, performing operations over the same or similar datasets becomes very easy using its programming language interface. The examples of this book are programmed in such a manner that anyone can replicate and practice the same. This makes it a useful teaching tool in learning R for machine learning, thus enhancing the knowledge of data mining schemes.

4.1.3 Matlab

Matlab combines the analysis and design environment in a single desktop. It mainly works on the data stored in matrixes and arrays. It is a multi-paradigm environment which provides the ease of numerical computing and a research community. It is developed by MathWorks. It provides the facility of data manipulations, plotting functions, algorithms implementation, and interface management. Programs written in other languages (C, C++, Java, C#) are also understood easily by Matlab. Matlab was meant for numerical computing itself, but the optional toolbox known as MuPAD symbolic engine is also available for symbolic computing. Simulink is the package of Matlab used for graphics. As of now, more than 4 million users are using Matlab.

4.1.4 Weka

Weka is an open source software which is developed for executing machine learning algorithms. Weka contains a variety of tools for classification, regression, association, and clustering. It is licensed under GNU. Initially, Weka was developed for handling agricultural data; later in 1997, it was completely enhanced to support machine learning algorithms. It can also handle multiple file formats such as CSV. The necessary tools are already installed in Weka. The facility of the package manager is also available for installing the required packages and uninstalling the useless packages. Weka is developed in the Waikato University of New Zealand where Waikato Environment for Knowledge Analysis (Weka) is programmed in Java. It is executable on almost every platform and is mainly designed to help researchers.

4.2 Experiments

We prefer using R language for experimenting with machine learning models. Rattle and RStudio are both built on the statistical R. Both the tools can be easily understood without having a deep knowledge of R. Rattle is more simple and easy to learn. Rattle develops the project with a file which is processed and navigated through the different phases of executing the machine learning model. R is a powerful language meant not only for machine learning but also for other data mining tools. Rattle makes the execution of the machine learning model just a click away. It works by calling the in build functions itself, which removes the tedious job of a programmer. It is one of the concepts of optimization in machine learning to execute the models. This can be done easily with GUI such as Rattle and Rstudio. It saves the current state of the task which can be named as the project. The project can be loaded anytime and can be modified at any stage. The transparency in the projects can lead to the success of science because of the repetition of executing requirements. Rattle is sufficient in terms of learning a user's needs according to the real-time application. It works as a stepping stone for learning the basics of machine learning. The layout of Rattle for the new project is presented in Figure 4.1. Running the project in Rattle follows the steps as discussed below:

1. Loading the dataset.

2. Setting the target variable among all the variables.

3. Exploring different categories of each variable.

4. Testing the dataset to check whether it is distributed or not.

5. Transforming data to make it suitable for a single scale.

6. Selecting suitable models.

7. Building the models.

8. Evaluating the models with different datasets for results.

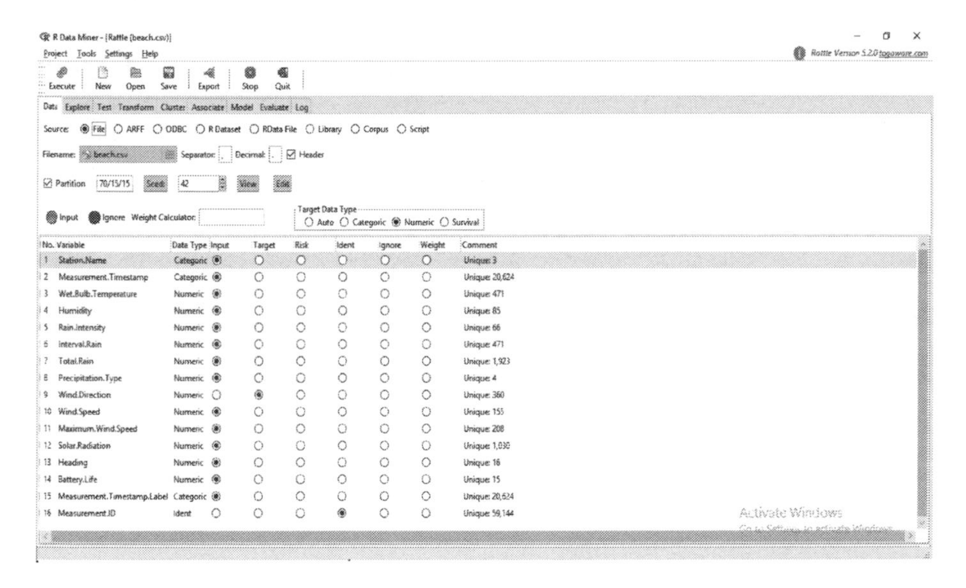

FIGURE 4.1: Graphical representation of Rattle: R statistical tool

Why we choose Rattle?
Rattle GUI is a free software which allows anyone to download and execute it. We can also call it an open-source software as it provides the source code as well. There is no restriction or license requirement for the same. The front page of the January 7th, 2009 *The New York Times* noted that it is really useful for high-end analysts with readily available code. R is not only free and open software but also reviewed software. Many bugs were fixed over the years, so the chances of failure are less. Similarly, if talking about RStudio, it is the tool used for programming. RStudio is an open-source GUI tool for the R language. It is preferred the most by the data mining project developers. It was developed by Joseph J. Allaire. He is also the inventor of ColdFusion Markup Language (CFML). RStudio is not ready to code, every single code has to be written. The layout of RStudio is shown in Figure 4.2. The various steps for building the Rstudio file are discussed below:

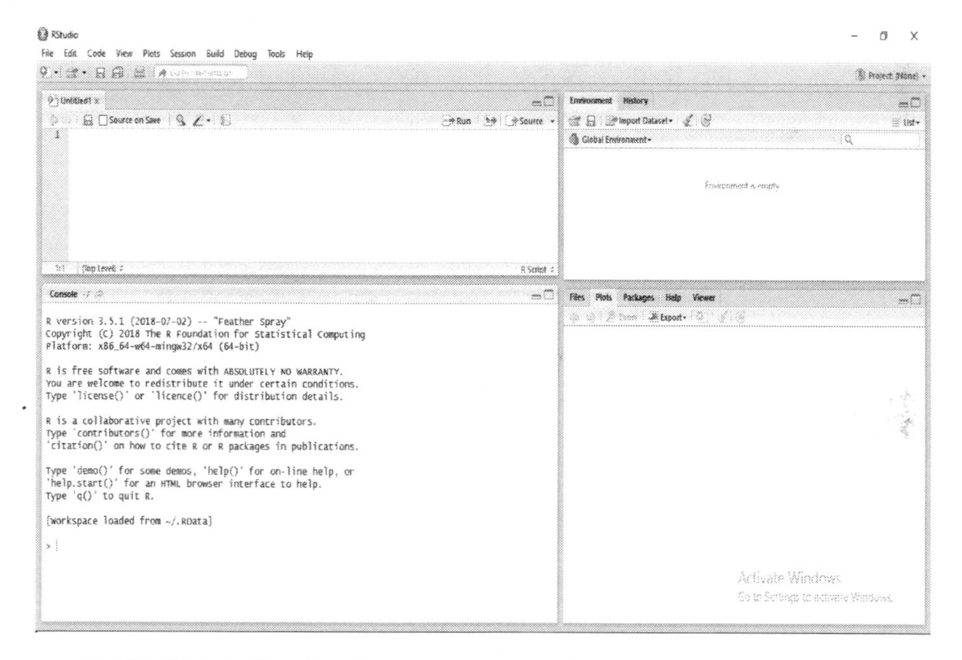

FIGURE 4.2: Graphical representation of RStudio: R statistical tool

1. Open File option and click on New File.

2. Write a code, specifically mentioning the steps required to build a machine learning model as below:

    - Include the R libraries required to execute the program.
    - Define the required variables.
    - Explicitly mention the training and testing ratio.

- Load the dataset file.
- Examine the variables present in the file.
- Set the target variable.
- Select the training and testing sets.
- Build the machine learning model.
- Compare the predicted values with the actual values.
- Compute the evaluation parameters.

R programming can be done in Ratte and RStudio. There are some advantages of using R mentioned as below:

- It is highly preferred by researchers for statistical analysis.
- It incorporates the code to perform statistical tests and models.
- R is built by highly qualified and experienced statisticians.
- The graphics library is coded in such a way that it enhances the results' visibility.
- R is an open-source software validated under the governance and documented for the U.S. Food and Drug Administration.
- R is licensed under the GNU General Public License, with the copyright held by the R Foundation for Statistical Computing.
- R is a free software, so it does not require any license.
- Anyone can provide code enhancements, and fix the bugs any time and from anywhere.
- More than 4800 packages are available by multiple repositories.
- It is executable on any Windows and on any processor.
- It supports many file formats such as .csv, .sas, and many more.
- R has active groups to discuss the queries and find the solutions.
- Books are also available on R, which contains a good collection of libraries (the Springer Use R! series).

4.2.1 Dataset

The machine learning models help to efficiently maintain IoT data and improve the performance of IoT devices. To understand each concept and working of machine learning models, we have downloaded the IoT sensors dataset. This dataset contains the weather sensor data maintained by the Chicago Park District at beaches along Chicago's Lake Michigan lakefront. For better practice, readers can download the same dataset from this link:
"https://data.world/cityofchicago/beach-weather-stations-automated-sensors".
This dataset contains 18 columns, namely, Station Name Measurement, Timestamp, Air Temperature, Wet Bulb Temperature, Humidity, Rain Intensity, Interval Rain, Total Rain, Precipitation Type, Wind Direction, Wind Speed, Maximum Wind Speed, Barometric Pressure, Solar Radiation, Heading, Battery Life, Measurement Timestamp Label, and Measurement ID. The two columns namely, Air Temperature and Barometric Pressure do not have any data as shown in Figure 4.3, so remove them before the experiments.

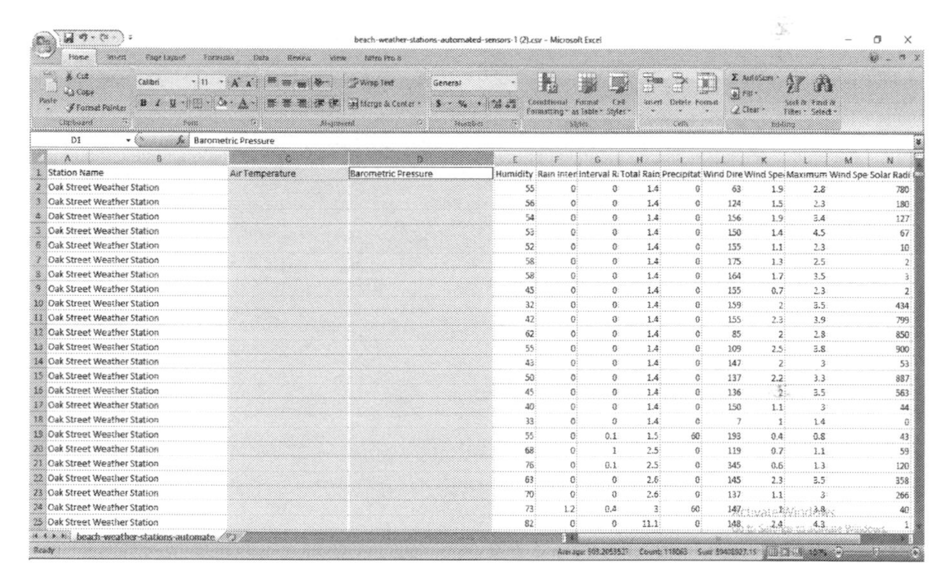

Station Name	Air Temperature	Barometric Pressure	Humidity	Rain Inter	Interval R	Total Rain	Precipitat	Wind Dire	Wind Spe	Maximum Wind Spe	Solar Radi
Oak Street Weather Station			55	0	0	1.4	0	63	1.9	2.8	780
Oak Street Weather Station			56	0	0	1.4	0	124	1.5	2.3	180
Oak Street Weather Station			54	0	0	1.4	0	156	1.9	3.4	127
Oak Street Weather Station			53	0	0	1.4	0	150	1.4	4.5	67
Oak Street Weather Station			52	0	0	1.4	0	155	1.1	2.3	10
Oak Street Weather Station			58	0	0	1.4	0	175	1.3	2.5	2
Oak Street Weather Station			58	0	0	1.4	0	164	1.7	3.5	3
Oak Street Weather Station			45	0	0	1.4	0	155	0.7	2.3	2
Oak Street Weather Station			32	0	0	1.4	0	159	2	3.5	434
Oak Street Weather Station			42	0	0	1.4	0	155	2.3	3.9	799
Oak Street Weather Station			62	0	0	1.4	0	85	2	2.8	850
Oak Street Weather Station			55	0	0	1.4	0	109	2.5	3.8	900
Oak Street Weather Station			43	0	0	1.4	0	147	2	3	53
Oak Street Weather Station			50	0	0	1.4	0	137	2.2	3.3	887
Oak Street Weather Station			45	0	0	1.4	0	136	2	3.5	563
Oak Street Weather Station			40	0	0	1.4	0	150	1.1	3	44
Oak Street Weather Station			33	0	0	1.4	0	7	1	1.4	0
Oak Street Weather Station			55	0	0.1	1.5	60	193	0.4	0.8	43
Oak Street Weather Station			68	0	1	2.5	0	119	0.7	1.1	59
Oak Street Weather Station			76	0	0.1	2.5	0	345	0.6	1.3	120
Oak Street Weather Station			63	0	0	2.6	0	145	2.3	3.5	358
Oak Street Weather Station			70	0	0	2.6	0	137	1.1	3	266
Oak Street Weather Station			73	1.2	0.4	3	60	147	1	3.8	40
Oak Street Weather Station			82	0	0	11.1	0	148	2.4	4.3	1

FIGURE 4.3: Missing values in dataset in the experimental IoT dataset

4.3 Supervised Learning

This problem refers to the learning in which the features are measured to compute the final outcome. All the previous examples belong to this class of learning. The unit of measurement for the final output depends upon the features (acts as an input). The function is defined within the algorithm which maps the available input into the required output. This kind of learning is mainly preferred in data mining tasks. The algorithms are defined in such a manner that the data is predicted based upon the inputs and connected signal. The signal may come from the environment to activate the system. The unseen instances are predicted correctly with the algorithm. For example, setting the admission criteria for the play school. The scenario is fixed to form the training data as discussed in the Table below.

TABLE 4.1: Classification of toddlers

Name	Age	Height	Abilities
Sahil	2.5 yrs	2'1"	Can speak Hindi
Rahul	2.3 yrs	2'5"	Can draw sleeping and standing lines
Siya	2.8 yrs	3'2"	Very talkative
Divya	2.4 yrs	2'8"	Can speak Hindi, English and Punjabi
Kajal	3 yrs	2'5"	Cannot speak clearly

Now think over this table and set the basic eligibility criteria. The admission in a play-way school requires the student to fulfill the minimum age of 2.5 yrs. On the basis of which, Sahil, Siya, and Kajal are eligible. It is a small list which can be manually traced, but for a long list, we require the computerized inspection. Table 4.1 acts as the training data, and the student is admissible or not is the response. The rule defined for this example is:

$$if(age >= 2.5), \tag{4.1}$$

then the student is eligible.

In supervised learning, the training data is always present, and output is the response from it. The constraint to be kept in mind while designing a supervised learning model/algorithm is the complexity and biasness. The complexity refers to the function complexity the machine is going to learn. The function defined in Eq. 4.1 should consider space complexity and time complexity. Biasness refers to the inaccurate learning of the system due to the problem of overfitting and underfitting.

Exercise 1: The data is the collection of websites present at the Microsoft website (https://www.microsoft.com/en-us/research/project/mslr). By just viewing the data, try to define the rule which differentiates between normal and spam websites.

4.3.1 Unsupervised Learning

It is the type of learning in which hidden insights of data are focused and the output is defined in terms of patterns detected in the input. The training data is not present and the response is also not definite. For example, considering the same example, we went to a park and met five toddlers, but we don't have actual information of all. Now, without this information, we may decide to find the most intelligent one, by interacting more with them. So, the hidden information needs to be detected. Exploratory analysis is one of its main applications as unsupervised learning detects the structure automatically. It is impossible to generate new trends for the data without knowing the insights of data. Unsupervised learning helps to observe the hidden patterns and work more deeply than the supervised learning. In supervised learning, it is easy to judge the performance of a model with the help of evaluation results, such as accuracy, root mean square error, and so on. However, in unsupervised learning, there is no universal measure basis for comparison.

4.4 Classification

As the name suggests, classification means to classify, so that the prediction of the numerical data becomes easy. It is one of the basic modes of supervised learning. In this, the objective is to train the system in such a manner that the system learns the different hidden patterns of the data. Let us understand it with a real-world scenario. Suppose you design a cognitive search engine which detects spam web pages automatically. How does the system get to know that the page is spam? The system may check the repetition of keywords, can verify the validation of images with the content of the website, checking the authoritative websites, inspecting the quality of in-links of the web page, and many more spam characteristics. But to apply every spam detection algorithm is not possible in practice. Here, the machine learning with classification comes into the role when there is a definite answer to the problem. So, either the page is spam or not. The machine learning algorithm takes the quality of a web page, which can take two possible values, i.e., spam or non-spam. This kind of classification is binary classification. The well-designed spam detection system is able to detect the spam web pages, but the question is, how much spam is injected into the web page? Sometimes, the system trained by us considers parameters which actually do not exist. Such kind of learning is known as outlier learning. This learning is discussed in the coming chapters. What do we actually need to make sure the result is not bias? As we target to find the spam injection, so the spam score works in this consideration. The features such as in-links, keywords, which are responsible for manipulations of the web page, are being injected in terms of spam. This score can make it easy to understand how much the web page is infected. Now, the spam score lies in the range from 1 to 100, any numerical value. This is another kind of classification, known as multi-class classification, in which a target variable can take a definite numerical value. Considering the IoT dataset discussed earlier, the summary of the dataset is presented in Figure 4.4. This figure computes the various parameters of each feature. The sixteen features are presented with their minimum value, 1st Quadrant, Median, Mean, 3rd Quadrant, and maximum value. The wind direction and the heading are the variables lying with the multi-class classification constraint.

```
                  Station.Name    Measurement.Timestamp  Wet.Bulb.Temperature
63rd Street Weather Station:19708  1/0/1900 0:00  :   3   Min.    :-20.10
Foster Weather Station     :20299  1/1/2016 0:00  :   3   1st Qu.:  0.00
Oak Street Weather Station :19137  1/1/2016 1:00  :   3   Median :  3.70
                                   1/1/2016 10:00:    3   Mean   :  6.99
                                   1/1/2016 11:00:    3   3rd Qu.: 15.40
                                   1/1/2016 12:00:    3   Max.   : 28.40
                                   (Other)        :59126
   Humidity       Rain.Intensity      Interval.Rain       Total.Rain      Precipitation.Type
Min.   :  0.00  Min.   :  0.0000   Min.   : 0.0000   Min.   :   0.0   Min.   : 0.000
1st Qu.: 59.00  1st Qu.:  0.0000   1st Qu.: 0.0000   1st Qu.:   0.0   1st Qu.: 0.000
Median : 71.00  Median :  0.0000   Median : 0.0000   Median :  14.3   Median : 0.000
Mean   : 69.46  Mean   :  0.1232   Mean   : 0.1753   Mean   : 121.8   Mean   : 3.055
3rd Qu.: 82.00  3rd Qu.:  0.0000   3rd Qu.: 0.0000   3rd Qu.: 190.2   3rd Qu.: 0.000
Max.   :100.00  Max.   :183.6000   Max.   :63.4200   Max.   :1056.1   Max.   :70.000

Wind.Direction   Wind.Speed       Maximum.Wind.Speed Solar.Radiation     Heading
Min.   :  0    Min.   :  0.00   Min.   :  0.000   Min.   :  -6.0   Min.   :  0.0
1st Qu.: 86    1st Qu.:  1.40   1st Qu.:  2.400   1st Qu.:   0.0   1st Qu.:  0.0
Median :196    Median :  2.50   Median :  3.800   Median :   3.0   Median :  1.0
Mean   :179    Mean   :  2.97   Mean   :  4.484   Mean   : 108.5   Mean   :122.3
3rd Qu.:271    3rd Qu.:  4.00   3rd Qu.:  5.900   3rd Qu.:  96.0   3rd Qu.:354.0
Max.   :359    Max.   :999.90   Max.   :999.900   Max.   :1277.0   Max.   :359.0

 Battery.Life    Measurement.Timestamp.Label                                  Measurement.ID
Min.   : 0.00   1/0/1900 0:00  :    3        63rdStreetWeatherStation            :    1
1st Qu.:11.90   1/1/2016 0:00  :    3        63rdStreetWeatherStation201504250900:    1
Median :12.10   1/1/2016 1:00  :    3        63rdStreetWeatherStation201504300500:    1
Mean   :13.05   1/1/2016 10:00:     3        63rdStreetWeatherStation201506220900:    1
3rd Qu.:15.10   1/1/2016 11:00:     3        63rdStreetWeatherStation201506221000:    1
Max.   :15.30   1/1/2016 12:00:     3        63rdStreetWeatherStation201506221100:    1
                (Other)        :59126        (Other)                             :59138
>
```

FIGURE 4.4: Summary of the experimental IoT dataset

4.5 Regression

Regression is the second case of supervised learning after classification. In this kind of learning, the target variable can take any decimal value which was a numerical value in multi-class classification. For example, the baby's birth at the time of birth is predicted by the ultrasound, which considers various features such as head circumference, femur length. The machine algorithm with regression technique is used for the same. Weather forecasting for the prediction of rain is a regression problem. There exist many machine learning models for regression. The most popular of these are simple linear regression model, lasso regression, multiple regression algorithm, logistic regression, and multivariate regression algorithms.

4.6 Clustering

Clustering is the problem of finding the hidden patterns of data. It is one of the unsupervised learning paradigms in which the presence of correlated data values is gathered together to form a block, known as 'cluster'. It indicates the absence of data labels to measure other parameters. Many clustering procedures are programmed, differentiating between the nature of the cluster. Clustering is the application to be applied to a diverse large dataset but with the consideration of supervising the computational complexity. In the broad sense, we can say, a cluster is a subgroup of data having similar characteristics. Group formation of different subgroups can be done on the basis of various parameters. The knowledge discovery, pattern extraction, and decision making are the specialization of an analyst who is developing clustering procedures. Indeed, the fact is data may belong to different domains, multiple scales, and diverse classes. The solution is the grouping of similar instances. For instance, there is a collection of n websites, having f features. The ranking module is expecting similar websites to crawl together to provide the quality score. The websites belong to different fields, different domains, and even different content size. Here, clustering is required because of the absence of data labels. The clustering algorithm finds the different clusters, wherein each cluster is computed having the same attributes. Clustering is also called simplification. K-means clustering is one of the best clustering techniques. It has experimented on the same dataset in Rattle function. The results are shown in Figure 4.5.

4.7 Summary and What's Next?

In this chapter, the different tools for implementing machine learning were highlighted. They include Python, Matlab, R, Weka. R has been a most powerful tool which is used in the examples covered in the next section. The complete basic programming concepts of R were elaborated on for better understanding of high-level programming, required for the development of a machine learning model. Various machine learning paradigms, were also explained, such as supervised learning, classification, regression, clustering, and so on.

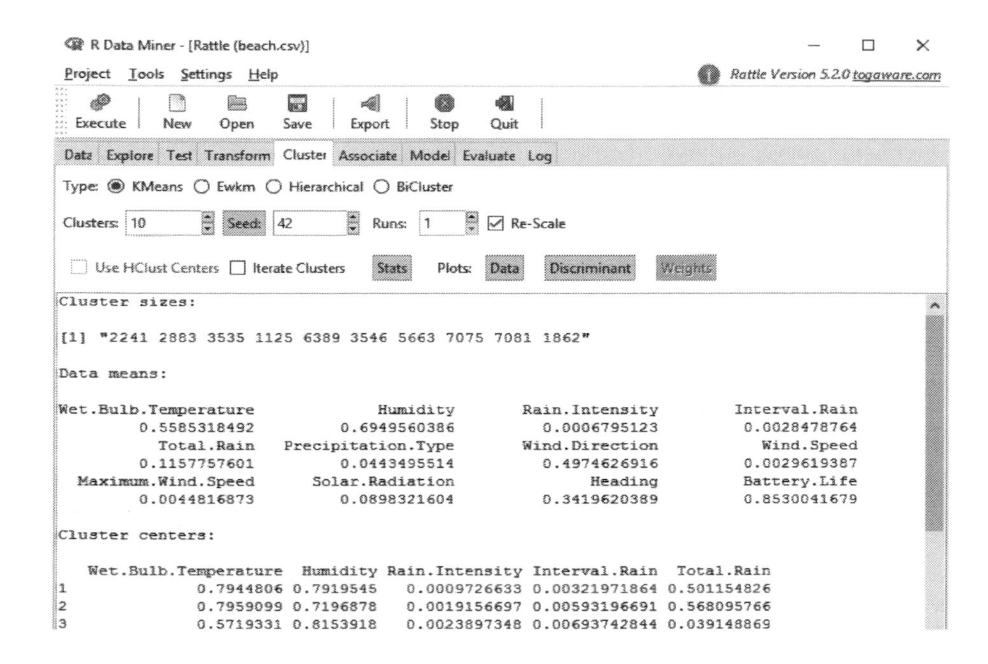

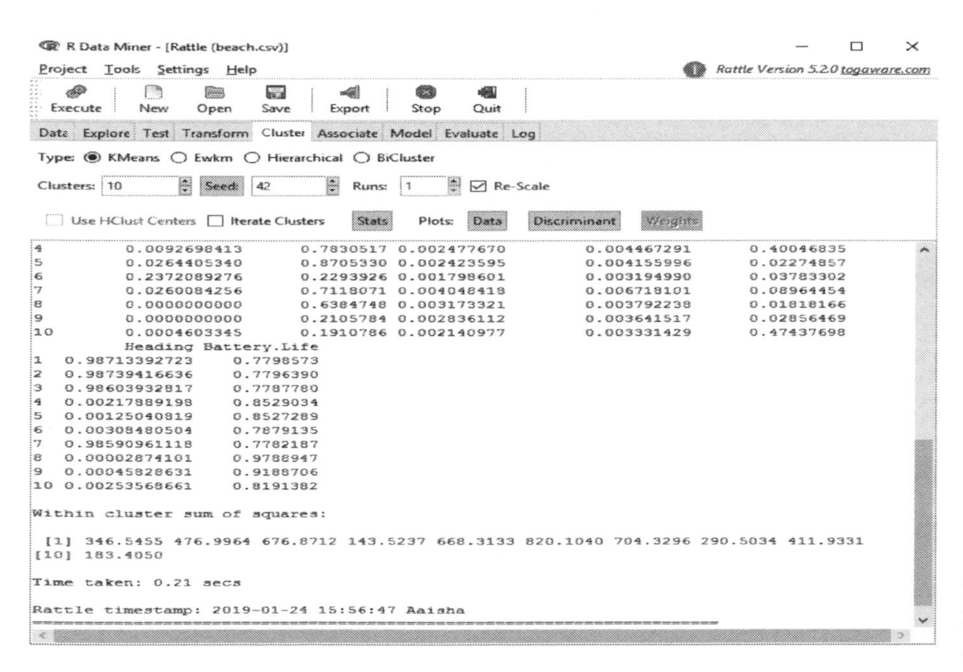

FIGURE 4.5: Clusters formed in k-means clustering

4.8 Exercises

Questions

1. Define the learning mechanism of machine learning.
2. List the different factors that affect the performance of machine learning.
3. What are the different tools to implement machine learning?
4. What is Rattle?
5. What do you mean by supervised learning?
6. What is unsupervised learning?
7. What is classification?
8. What is regression?
9. Define clustering.
10. What are the steps involved for building the RStudio file?

Multiple Choice Questions

1: What differentiates between the various structures such as vector, list, matrix, images, and videos?

(a) Memory
(b) Data size
(c) Data type
(d) All of these

2: Predicting the human body temperature is a:

(a) Binary classification problem
(b) Multi classification problem
(c) Regression problem
(d) None of them

3: Which tool is considered best for implementing a machine-learning model?

(a) Python **(b)** R

(c) Matlab **(d)** Weka

4: Which R tool by default includes the definition of in-build functions?

(a) Rattle **(b)** RStudio

(c) Weka **(d)** None of the above

5: Which tool contains the Scikit-Learn library of built-in functions?

(a) Python **(b)** R

(c) Matlab **(d)** Weka

6: MuPAD toolbox is used for which kind of processing?

(a) Machine learning modeling **(b)** Numerical computing

(c) Graphics representation **(d)** None of the above

7: Which platform supports Weka software?

(a) Windows **(b)** Linux

(c) Mac **(d)** All of the above

8: Who is the invented of the R statistical tool, i.e., Rattle?

(a) Guido van Rossum

(b) Joseph J. Allaire

(c) Ross Ihaka

(d) None of them

9: The system is trained with the training data to predict the testing data. This approach is followed by which kind of learning?

(a) Numerical learning

(b) Supervised learning

(c) Unsupervised learning

(d) Clustering

10: Which approach is implemented to find the hidden patterns of the data?

(a) Supervised learning

(b) Clustering

(c) Unsupervised learning

(d) All of above

11: Which learning learns the data beyond the basic parameters of indexed data?

(a) Numerical learning

(b) Outlier learning

(c) Decision learning

(d) Regression

12: The target value predicted with the scheme of regression value holds:

(a) Numerical value

(b) Decimal value

(c) Infinite value

(d) Character value

13: Which tool is developed at Waikato University, New Zealand?

(a) Python (b) R

(c) Matlab (d) Weka

14: Which tool helps in optimizing the results of machine learning models?

(a) Weka (b) R

(c) Matlab (d) All of above

15: Which tool enhances the security of credit cards?

(a) IoT (b) AI

(c) Machine learning (d) None of the above

16: List is the collection of parameters of:

(a) Single type (b) Multiple types

(c) Character type (d) None of the above

17: Which data type is preferred for learning from different units of data?

(a) Vectors (b) Lists

(c) Matrices (d) None of the above

18: giff file refers to which data?

(**a**) Image
(**b**) Video
(**c**) Audio
(**d**) None of the above

19: Multi-classification is adopted in case of:

(**a**) Multiple target values
(**b**) Multiple conditional values
(**c**) Multiple character values
(**d**) None of the above

20: The correlated data in a block is known as:

(**a**) Vector
(**b**) List
(**c**) Cluster
(**d**) Outliers

5

R Programming

5.1 Introduction

In recent years, the way we evaluate information has been changed dramatically. The sheer quantity of information we have at our disposal has risen enormously with the introduction of personal computers and the Internet. Companies have terabytes of information about the customers they communicate with, and organizations of government, academia, and private study have comprehensive archival and survey information on all study topics. New statistical methods were created by scholarly scientists before the invention of personal computers and the Internet. It takes years for programmers to adopt these techniques and incorporate them into the statistical packages that are commonly accessible to data analysts today.

Personal computers' emergence had another impact on how we analyze information. On mainframe computers, when data analysis was performed, computer time was valuable and unable to judge. Analysts would set up a laptop running with all the necessary parameters and choices. The resulting output may be dozens or hundreds of pages long when the operation ran. Modern data analysis has moved to a distinct paradigm with inexpensive and simple access provided by personal computers. Modern methods become extremely interactive rather than setting up a full data analysis at once, with the output from each phase serving as the input for the next level.

The emergence of personal computers (particularly the availability of high-resolution monitors) also influenced the understanding and presentation of outcomes. An image can really be worth a thousand words, and people are capable of extracting helpful data from visual presentations. Increasingly, modern information analysis depends on graphical presentations to discover significance and transmit outcomes. Today's data analysts need to access data from a wide variety of sources, merge, clean and annotate data pieces, analyze them using the latest methods, present findings in meaningful and graphically appealing ways, and incorporate the results into attractive reports that can be distributed to stakeholders and the public. Here, to perform all these computations involved in machine learning, we would be using the R language.

5.1.1 Basis of the R programming

Before going into the deep code of machine learning in the R, let us introduce the basic programming concepts of R. R was developed for the purpose of statistical analysis. The graphic support is established with the help of the graphics package. R is a handful programming language for researchers. All the programming features, such as predefined functions, libraries, packages, and keywords are supported by the R language. Our primary motive is to abstract the capabilities of R as the statistics toolbox while concentrating on all the prospective of the basic agenda which is required by a good programmer It is an open-source data analysis solution endorsed by a big and active research community around the world. We adopted the R language because of the following features.

- R is an extensive statistical platform offering all kinds of methods for data analysis.
- R includes sophisticated, not yet accessible statistical routines in other packages. In reality, on a weekly basis, fresh techniques are accessible for download. If you are a SAS user, imagine every few days getting a fresh SAS PROC.
- Most of the trading statistical software platforms cost thousands of bucks, if not tens of thousands. R is free of charge! The advantages are evident if you're a researcher or a student.
- R has state-of-the-art skills in graphics. It has the most extensive and strong function set available if you want to visualize complicated information.
- Getting information from various sources into a usable form can be a difficult proposal. R can readily import information from a broad range of sources, including text files, database management systems, statistical packages, and specific shops of information. It is also able to write information about these schemes. R can directly access information from websites, social media sites, and a broad variety of Internet information facilities.
- R features can be incorporated into other language written apps, including C++, Java, Python, PHP, Pentaho, SAS, and SPSS. This enables you to continue working in a language you may be acquainted with, while adding to your apps the capacities of R.
- R offers an unparalleled platform for simple programming of fresh statistical methods. It is readily expandable and offers a natural language for newly published techniques to be rapidly programmed.
- R operates on a broad range of systems including Windows, Unix, and

Mac OS X. It can probably run on any desktop you may have. If you don't want to learn a fresh language, there are a range of graphical user interfaces (GUIs) that offer R's authority through menus and dialogs.

5.1.2 Installing R

R is freely available at https:/ cran.r-project.org from the Comprehensive R Archive Network (CRAN). For Linux, Mac OS X, and Windows, precompiled binaries are accessible. Follow the instructions on the platform of your choice to install the base item. Later we're going to speak about adding features through optional package modules (also available from CRAN).

5.1.3 Working in R

R is an interpreted, case-sensitive language. At the command prompt (>) you can enter orders one at a time or execute a collection of instructions from a source file. There is a broad range of information, including vectors, matrices, information frames (like datasets), and lists (object collections). Most functionality can be given by built-in and user-created features as well as by object creation and manipulation. An item is anything to which a value can be allocated. For R, that's all we need to understand as a beginner. Each object has an attribute of a class that tells R how to manage it. During an interactive session, all items are stored in memory. If you are working in Windows, you can see the R command interface as shown in Figure 5.1.

5.2 Basic Commands

There are few R commands required to write a program in R. All you need to understand is the syntax and usage of each command. Below are the R commands.

5.2.1 Assignment

The most straightforward way to store a list of numbers is by using a c command unit (c stands for 'combine'). The concept is to store a list of numbers under a given name and use the name to refer to the information. The c command specifies a list, and the assignment is indicated with the symbol '<-.' Another word used to define the number list is to call it a vector. The numbers we write with the c command are separated by commas. For example, we create a variable with the name, marks, which contain the numbers, 23, 43, 25, 46, and 31. We can even generate the marks without using c command. Both the commands are executed and shown in Figure 5.2. The first command in

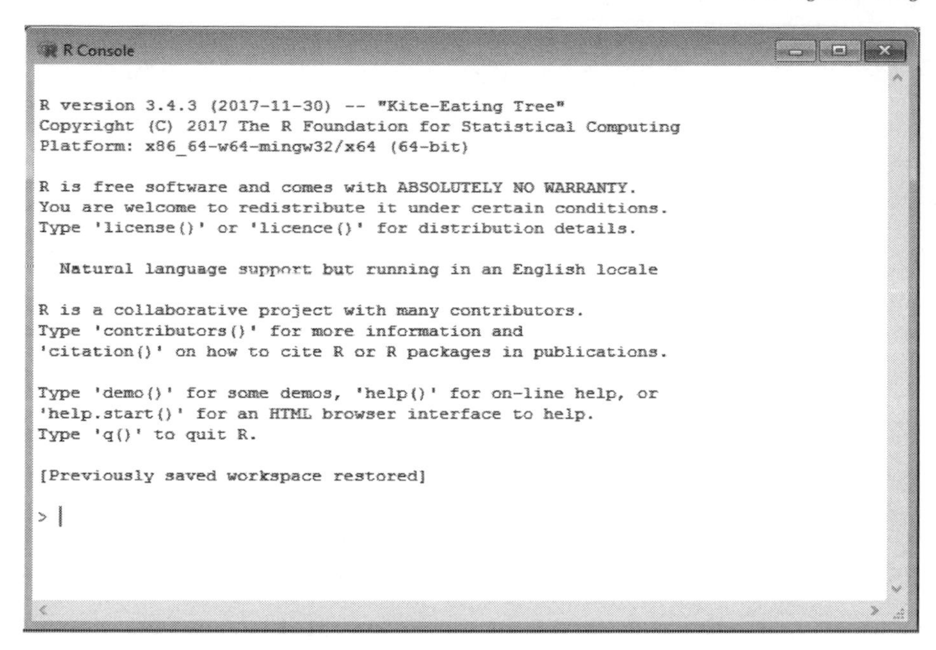

FIGURE 5.1: R interface in Windows operating system

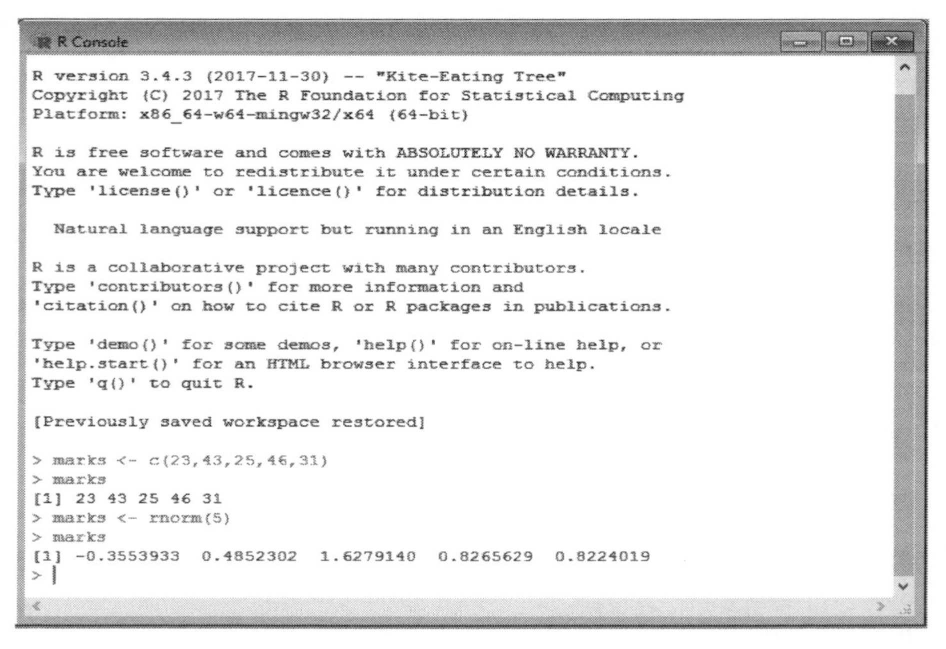

FIGURE 5.2: Functionality of assignment operator

the figure assigns the numbers to the variable marks. The second command creates a vector object named marks containing five random deviates from a standard normal distribution.

In the previous text and example, we have learned that how to assign the values to the variables. Assigning the strings is also very simple. As shown in Figure 5.3, the assignment is similar to that of numbers. Numbers are written within brackets, whereas strings are written within quotes. Printing the value of the string is done in two ways, either by just writing the name of the string or with the help of print command.

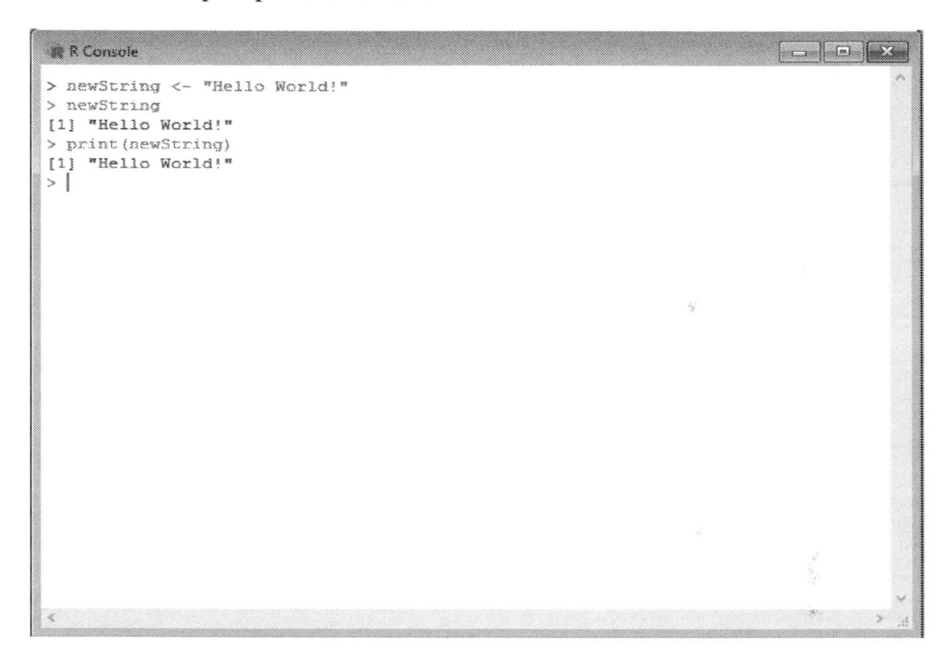

FIGURE 5.3: Functionality of string assignment

5.2.2 Comments

Comments are like assisting text in your R program and while executing your real program they are ignored by the interpreter. The single remark is written using '#' at the starting of the statement. R does not support multiple line comments. Any text scanned after this symbol would be ignored by the interpreter. Compilation and execution of comments are elaborated on in Figure 5.4.

5.3 Data Types

In a programming language, we need memory to store the data. This memory size depends upon the data type in which the variable/data is stored.

```
> newString <- "Hello World!"
> newString
[1] "Hello World!"
> print(newString)
[1] "Hello World!"
> # This is the first comment to understand the usage of commenting in R.
> # It wont print anything but let the user understand that how to increase the
> # readability of the program. Thank you!
> |
```

FIGURE 5.4: String assignment

5.3.1 Basic Data Types

5.3.1.1 Numbers

This data type is used to store a single value in a variable. The type of value is a real number. There is no need to mention the data type in R. The interpreter assigns the data type according to the value of a variable as shown in Figure 5.5.

5.3.1.2 String

R is not limited to numbers only; you can even store a character or a string in a variable. The same procedure is used with c function to store this type of data. Each value of a character or a string can be enclosed in a single quote or double quotes as shown in Figure 5.6. A single value can be accessed by an index which starts from one.

5.3.2 Structural Data Type

R has a broad range of data holding objects including scalars, vectors, matrices, arrays, information frames, and lists. They vary in terms of the sort of information they may hold, how they are generated, the complexity of their structure, and the notation used to define and access individual components.

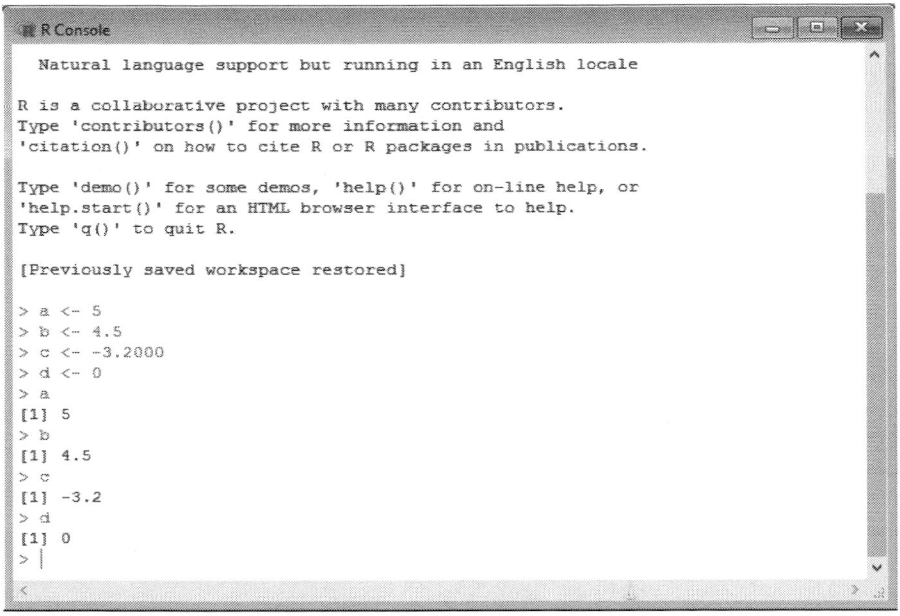

```
R Console
  Natural language support but running in an English locale

R is a collaborative project with many contributors.
Type 'contributors()' for more information and
'citation()' on how to cite R or R packages in publications.

Type 'demo()' for some demos, 'help()' for on-line help, or
'help.start()' for an HTML browser interface to help.
Type 'q()' to quit R.

[Previously saved workspace restored]

> a <- 5
> b <- 4.5
> c <- -3.2000
> d <- 0
> a
[1] 5
> b
[1] 4.5
> c
[1] -3.2
> d
[1] 0
> |
```

FIGURE 5.5: Working of number data type

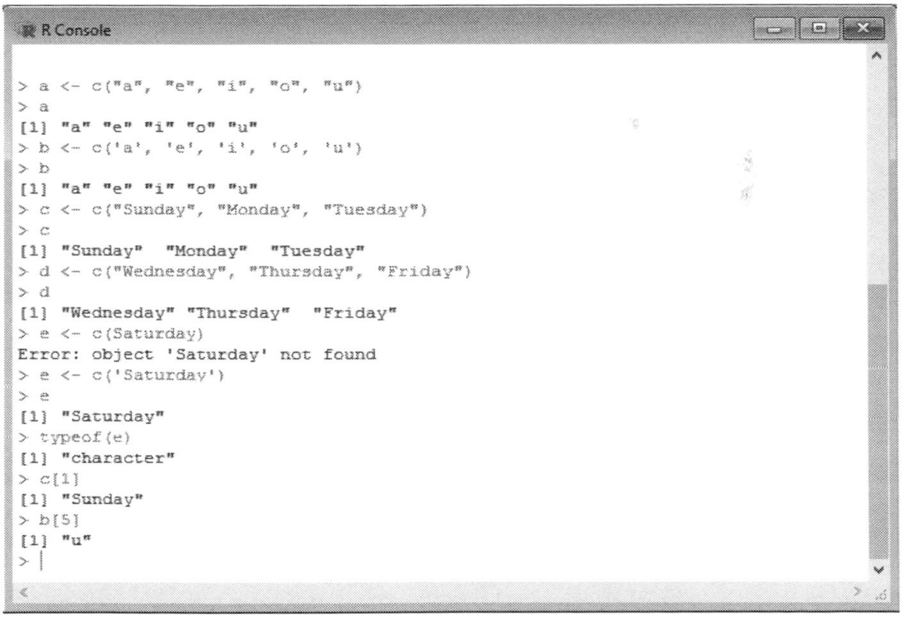

```
R Console

> a <- c("a", "e", "i", "o", "u")
> a
[1] "a" "e" "i" "o" "u"
> b <- c('a', 'e', 'i', 'o', 'u')
> b
[1] "a" "e" "i" "o" "u"
> c <- c("Sunday", "Monday", "Tuesday")
> c
[1] "Sunday"  "Monday"  "Tuesday"
> d <- c("Wednesday", "Thursday", "Friday")
> d
[1] "Wednesday" "Thursday"  "Friday"
> e <- c(Saturday)
Error: object 'Saturday' not found
> e <- c('Saturday')
> e
[1] "Saturday"
> typeof(e)
[1] "character"
> c[1]
[1] "Sunday"
> b[5]
[1] "u"
> |
```

FIGURE 5.6: Usage of string data type

Such a kind of data holder is categorized into the structural data type. Below are the different structural data holders.

5.3.2.1 Matrices

Matrix is the special case of an array which can hold the data up to two dimensions only. The data stored in this structure is uniform and scalar. A matrix() function of R is used to create the matrix. The syntax of this function is as follows

newmatrix<- matrix(vector, nrow=number_of_rows, ncol= number_of_columns, byrow=logical_value, dimnames=list(char_vector _rownames, char_vector_colnames)) The vector contains the data of the rows and columns, the nrow and the ncol specify the dimensions of the matrix and dimnames contains an optional row and column labels stored in character vectors. The option byrow indicates whether the matrix should be filled in by row (byrow=TRUE) or by column (byrow=FALSE). The default is by column. The matrix function is elaborated in Figure 5.7. The matrix transpose and multiplication are illustrated in Figure 5.8.

```
R Console
> newmatrix
     [,1] [,2] [,3] [,4]
[1,]    1    6   11   16
[2,]    2    7   12   17
[3,]    3    8   13   18
[4,]    4    9   14   19
[5,]    5   10   15   20
> newmatrix1 <- matrix(newvector, nrow=2, ncol=2, byrow=TRUE, dimnames=list(rnames, rcols))
> newmatrix1
   R2 C2
R1 12 24
C1 36 48
> newmatrix1 <- matrix(newvector, nrow=2, ncol=2, byrow=FALSE, dimnames=list(rnames, rcols))
> newmatrix1
   R2 C2
R1 12 36
C1 24 48
> newmatrix[2,]
[1]  2  7 12 17
> newmatrix[,2]
[1]  6  7  8  9 10
> newmatrix[2,2]
[1] 7
> newmatrix[1,c(2,4)]
[1]  6 16
> |
```

FIGURE 5.7: Matrix creation and matrix element retrieval

In this figure, first the new matrix is created using the command: newmatrix<- matrix(1:20, nrow=5, ncol=4). In this command, the random numbers from 1 to 20 are filled in 5 rows with 4 columns. Then the commands are executed for the creation of another matrix. newvector<- c(12, 24, 36,48)

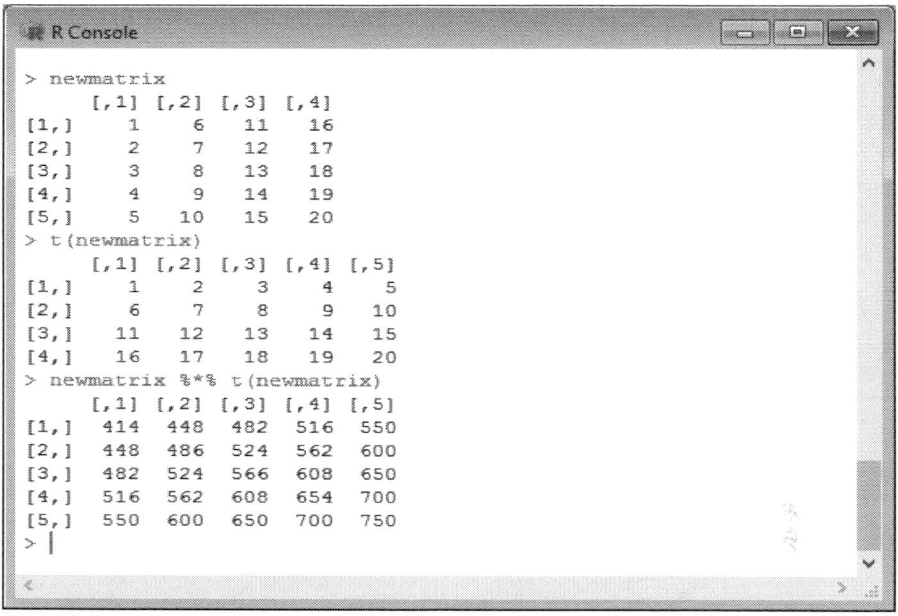

```
R Console
> newmatrix
     [,1] [,2] [,3] [,4]
[1,]    1    6   11   16
[2,]    2    7   12   17
[3,]    3    8   13   18
[4,]    4    9   14   19
[5,]    5   10   15   20
> t(newmatrix)
     [,1] [,2] [,3] [,4] [,5]
[1,]    1    2    3    4    5
[2,]    6    7    8    9   10
[3,]   11   12   13   14   15
[4,]   16   17   18   19   20
> newmatrix %*% t(newmatrix)
     [,1] [,2] [,3] [,4] [,5]
[1,]  414  448  482  516  550
[2,]  448  486  524  562  600
[3,]  482  524  566  608  650
[4,]  516  562  608  654  700
[5,]  550  600  650  700  750
> |
```

FIGURE 5.8: Matrix transpose and multiplication

is used to create a new vector, with the defined set of names. By default, the matrix is filled by column, which is manipulated with the two commands, where byrow= TRUE and byrow=FALSE. The retrieval of the matrix elements can be done with the help of the index. First, the second-row elements are retrieved, and then the second column is retrieved. Finally, the elements in the first row and the fourth and fifth columns are selected.

5.3.2.2 Arrays

An array is the combination of data placed uniformly in the dimensions. Unlike matrix, an array can be of any dimensions, i.e., 1, 2 or many more. There is a predefined function in R for the creation of arrays. Its syntax is as follows.

```
newarray<- array(vector, dimensions, dimnames)
```

In this, the data is listed in vector, the dimensions are the maximal index for each dimension, and dimnames is used to mention the dimension list. The dimnames is the optional parameter. The array creation operations are experimented and shown in Figure 5.9.

The retrieval can be done in a similar manner with the help of the index. The index contains the dimension list. It is shown in Figure 5.10.

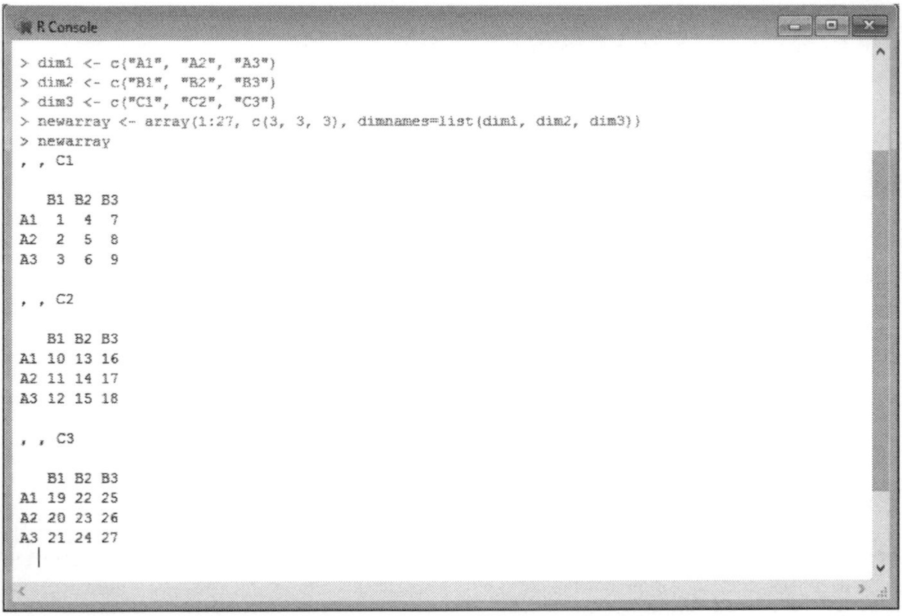

```
R Console
> dim1 <- c("A1", "A2", "A3")
> dim2 <- c("B1", "B2", "B3")
> dim3 <- c("C1", "C2", "C3")
> newarray <- array(1:27, c(3, 3, 3), dimnames=list(dim1, dim2, dim3))
> newarray
, , C1

   B1 B2 B3
A1  1  4  7
A2  2  5  8
A3  3  6  9

, , C2

   B1 B2 B3
A1 10 13 16
A2 11 14 17
A3 12 15 18

, , C3

   B1 B2 B3
A1 19 22 25
A2 20 23 26
A3 21 24 27
```

FIGURE 5.9: Illustration of array creation

```
R Console
> newarray[,,]
, , C1

   B1 B2 B3
A1  1  4  7
A2  2  5  8
A3  3  6  9

, , C2

   B1 B2 B3
A1 10 13 16
A2 11 14 17
A3 12 15 18

, , C3

   B1 B2 B3
A1 19 22 25
A2 20 23 26
A3 21 24 27

> newarray[1,1,1]
[1] 1
> newarray[2,,2]
B1 B2 B3
11 14 17
>
```

FIGURE 5.10: Function to retrieve array element

5.3.2.3 Data Frames

These are the structures which can hold data of different types. In other words, we can say, a data frame is a combination of multiple data-type variables, such as integer, character, or string. A data frame can be created with the inbuilt function, i.e., data.frame.
newdataframe<- data.frame(col1, col2, col3,...)

In this, the col1, col2, col3 are the column vectors. The data frame examples are displayed in Figure 5.11.

```
R Console
> student_id <- c(1023, 1024, 1025, 1027)
> student_name <- c("pooja", "john", "peter", "carley")
> student_marks <- c(98, 56, 76, 89)
> student_grade <- c('A', 'D', 'C', 'B')
> student_record <- data.frame( student_id, student_name, student_marks, student_grade)
> student_record
  student_id student_name student_marks student_grade
1       1023        pooja            98             A
2       1024         john            56             D
3       1025        peter            76             C
4       1027       carley            89             B
> student_record$student_name
[1] pooja  john   peter  carley
Levels: carley john peter pooja
> student_record[1:2]
  student_id student_name
1       1023        pooja
2       1024         john
3       1025        peter
4       1027       carley
> student_record[c("student_id", "student_grade")]
  student_id student_grade
1       1023             A
2       1024             D
3       1025             C
4       1027             B
> |
```

FIGURE 5.11: Example of data frame creation and element retrieval

In this figure, the student_record data frame is created. It consists of four columns, i.e., student_id of numeric type, student_name of string type, student_marks of numeric type, and student_grade of character type. It is created with the help of the data frame function. After creation, the elements are retrieved first with the help of a column, i.e., student_name. Second, they are retrieved with the index and third, with multiple columns retrieval.

5.3.2.4 Lists

This is the most complex structure as it can hold any data type, ranging from character, integers, arrays, and many more. The list function is used for creating this structure.

In Figure 5.12, different variables, such as an array, string, frame, and factor are combined to form a list. For two reasons, lists are significant R

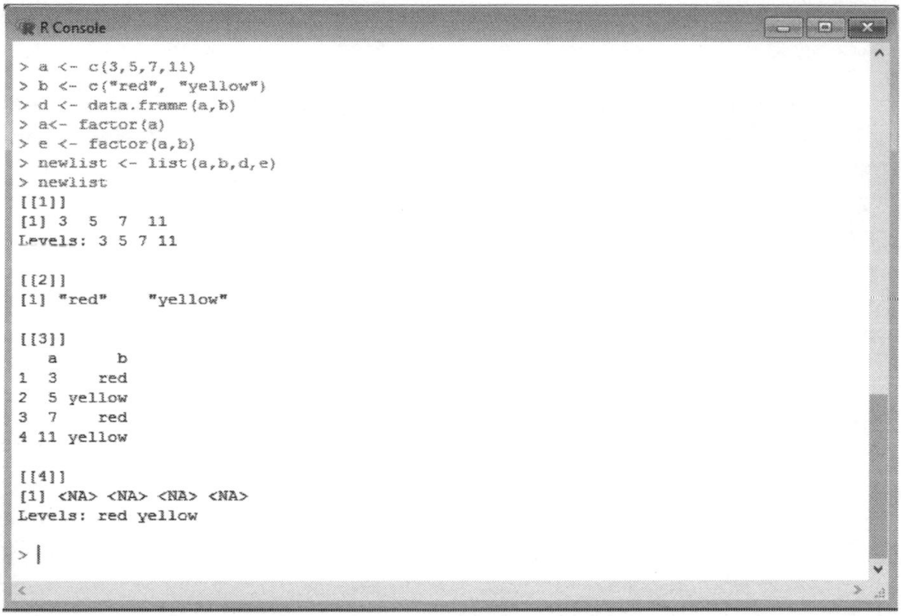

```
R Console
> a <- c(3,5,7,11)
> b <- c("red", "yellow")
> d <- data.frame(a,b)
> a<- factor(a)
> e <- factor(a,b)
> newlist <- list(a,b,d,e)
> newlist
[[1]]
[1] 3  5  7  11
Levels: 3 5 7 11

[[2]]
[1] "red"    "yellow"

[[3]]
   a      b
1  3    red
2  5 yellow
3  7    red
4 11 yellow

[[4]]
[1] <NA> <NA> <NA> <NA>
Levels: red yellow

>
```

FIGURE 5.12: Creation of list using data frame

structures. First, they enable you to simply organize and recall disparate data. Secondly, it list the outcomes of different R features. It is up to the analyst to pull out the necessary parts. In subsequent sections, you will see countless examples of features returning lists.

5.3.2.5 Factors

In computer languages, we can play with the variables to store the data. There are different variables, which play different roles. There are three variables: nominal, ordinal, and continuous. The variables without order are known as nominal variables. Fruits category is an example of a nominal variable. The variable with order is known as an ordinal variable, such as student marks. The variables within some range and order are known as continuous variables. An example of a continuous variable is the accuracy of the model. The R language considers nominal and ordinal variables, and these variables are known as factors.

In Figure 5.13, the function factor is used. For the nominal variables, factor function stores the value as a vector of integers. For the ordinal variables, the order parameter is added which takes the value of TRUE or FALSE.

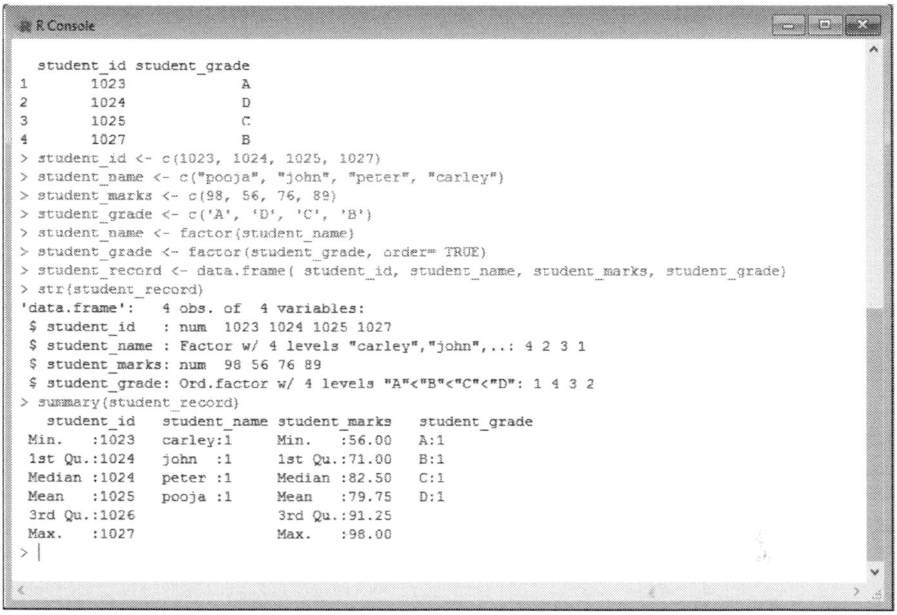

```
R Console
  student_id student_grade
1       1023             A
2       1024             D
3       1025             C
4       1027             B
> student_id <- c(1023, 1024, 1025, 1027)
> student_name <- c("pooja", "john", "peter", "carley")
> student_marks <- c(98, 56, 76, 89)
> student_grade <- c('A', 'D', 'C', 'B')
> student_name <- factor(student_name)
> student_grade <- factor(student_grade, order= TRUE)
> student_record <- data.frame( student_id, student_name, student_marks, student_grade)
> str(student_record)
'data.frame':   4 obs. of  4 variables:
 $ student_id   : num  1023 1024 1025 1027
 $ student_name : Factor w/ 4 levels "carley","john",..: 4 2 3 1
 $ student_marks: num  98 56 76 89
 $ student_grade: Ord.factor w/ 4 levels "A"<"B"<"C"<"D": 1 4 3 2
> summary(student_record)
   student_id    student_name student_marks   student_grade
 Min.   :1023   carley:1     Min.   :56.00   A:1
 1st Qu.:1024   john  :1     1st Qu.:71.00   B:1
 Median :1024   peter :1     Median :82.50   C:1
 Mean   :1025   pooja :1     Mean   :79.75   D:1
 3rd Qu.:1026                3rd Qu.:91.25
 Max.   :1027                Max.   :98.00
>
```

FIGURE 5.13: Factor creation and factor representation

5.3.2.6 Vectors

It is the special case of array data type, which can hold only one-dimensional data. The data in the vector can be numeric, character, string, or logical. You cannot group the multiple data types in a single vector. The retrieving of one or more elements is possible in vector. This is done with the help of the index. The index of the vector starts with one. The different vectors of creation and retrieval are discussed in Figure 5.14.

5.3.3 Operators

The operators are used to perform some computation on the variables holding data. These operators are either specific with the task or used for depicting the relationship between two or more variables. There are different types of operators in R. A few of the frequently used operators are discussed below.

5.3.3.1 Arithmetic Operators

These operators are used to perform mathematical operations, such as adding and multiplying. A list of accessible arithmetic operators of R is shown in Table 5.1. The experiments performed using arithmetic operators are illustrated in Figure 5.15.

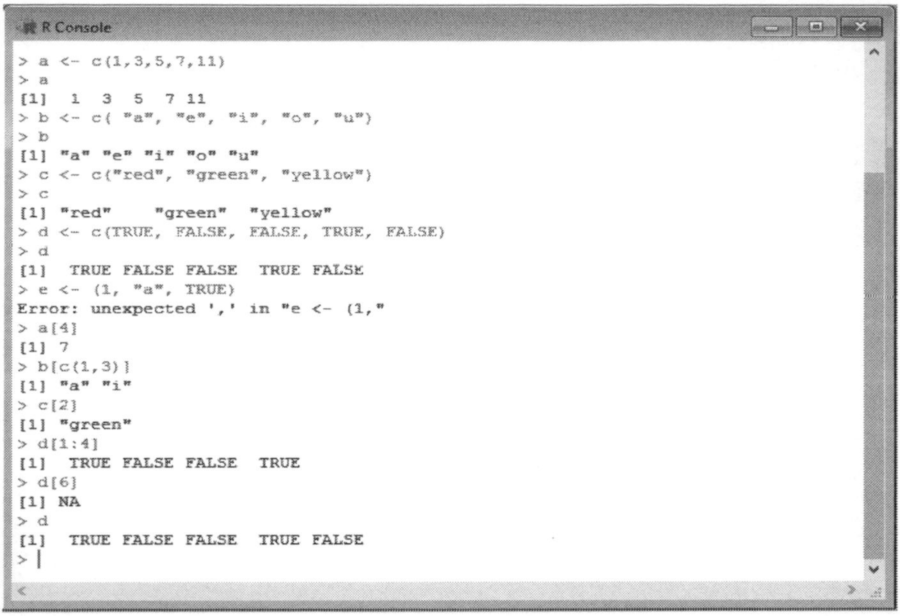

```
R Console
> a <- c(1,3,5,7,11)
> a
[1]  1  3  5  7 11
> b <- c( "a", "e", "i", "o", "u")
> b
[1] "a" "e" "i" "o" "u"
> c <- c("red", "green", "yellow")
> c
[1] "red"    "green"  "yellow"
> d <- c(TRUE, FALSE, FALSE, TRUE, FALSE)
> d
[1]  TRUE FALSE FALSE  TRUE FALSE
> e <- (1, "a", TRUE)
Error: unexpected ',' in "e <- (1,"
> a[4]
[1] 7
> b[c(1,3)]
[1] "a" "i"
> c[2]
[1] "green"
> d[1:4]
[1]  TRUE FALSE FALSE  TRUE
> d[6]
[1] NA
> d
[1]  TRUE FALSE FALSE  TRUE FALSE
> |
```

FIGURE 5.14: Vector creation and vector element retrieval

TABLE 5.1: Arithmetic operators

Operator	Description
+	Adding two or more numbers
-	Subtracting two or more numbers
*	Multiplying two or more numbers
/	Dividing two or more numbers
%%	Modulus operator to compute the remainder by division
%/%	Division to compute the quotient as integer
∧	Exponential computation

5.3.3.2 Relational Operators

These operators are used to draw a comparison between two or more variables. These operators are mainly used to test the condition to be used in the loops. Table 5.2 discusses the various relational operators in R. These operators are shown in Figure 5.16.

5.3.3.3 Logical Operators

This operator is valid only to vectors of type logical, number, or complex numbers. All figures greater than one are considered a logical value, i.e., TRUE. Here is the list of logical operators used by the R language in Table 5.3. These

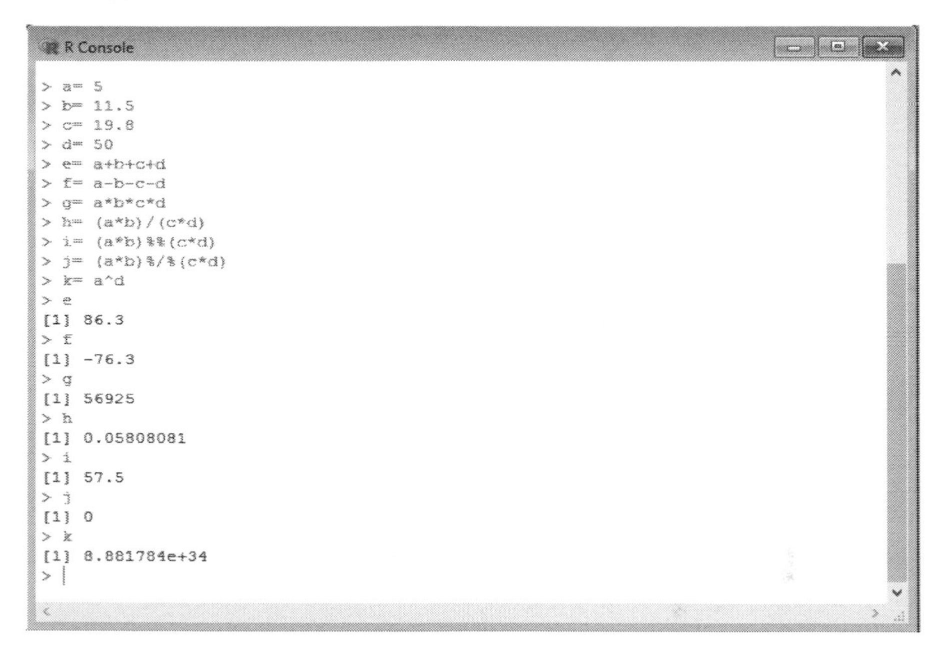

FIGURE 5.15: Examples of operations performed using arithmetic operators

TABLE 5.2: Relational operators

Operator	**Description**
<	This less than operator results in true if the variable is less than the other.
>	This greater than operator results in true if the variable is greater than the other.
<=	This less than equal to operator results in true if the variable is less than or equal to the other.
>=	This greater than equal to operator results in true if the variable is greater than or equal to the other.
==	This equal to operator results in true if both the variable are equal.
!=	This not equal to operator results in true if both the variable are not equal.

operations are possible only for numeric, logical, or complex types. The listed operators are shown in Figure 5.17.

5.3.4 Graphics

Graphics are important for conveying important features of the data. They can be used to examine marginal distributions, summarizing the large dataset

FIGURE 5.16: Examples of operations performed using relational operations

TABLE 5.3: Logical operators

<table>
<tr><th>Operator</th><th>Description</th></tr>
<tr><td>&</td><td>Element wise logical AND</td></tr>
<tr><td>|</td><td>Element wise logical OR</td></tr>
<tr><td>&&</td><td>Logical AND</td></tr>
<tr><td>!</td><td>Logical NOT</td></tr>
<tr><td>||</td><td>Logical OR</td></tr>
</table>

and for depicting the relation between the data features. plot() is the most common R function.

5.3.4.1 Histograms

Histograms can be easily created in the R using the hist() function and a vector of values. The parameters used by the hist() function are breaks, xlab, ylab, xlim, and ylim. The graphical representation is presented in Figure 5.18.

5.3.4.2 Scatter Diagrams

There are many ways to create a scatter plot in R. The basic function is plot(x, y), where x and y are numeric vectors denoting the (x,y) points to plot. This is created with the plot function as shown in Figure 5.19. The data can be sorted with the help of the sort function and presented with ‘type’ function.

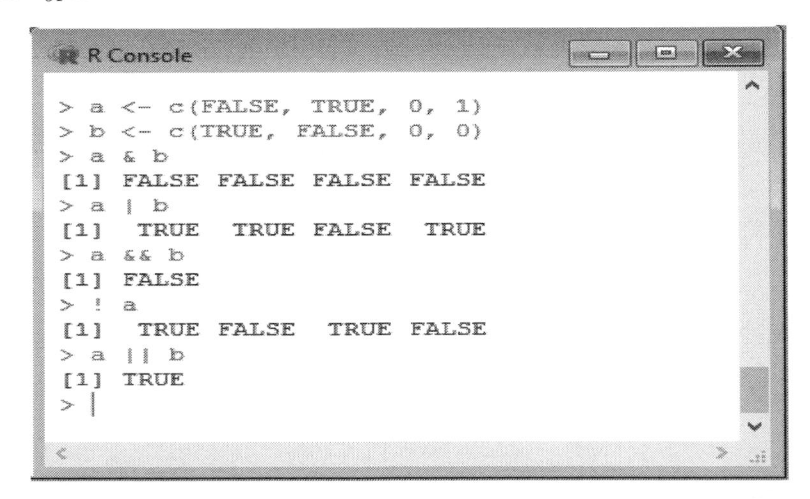

FIGURE 5.17: Examples of operations performed using logical operations

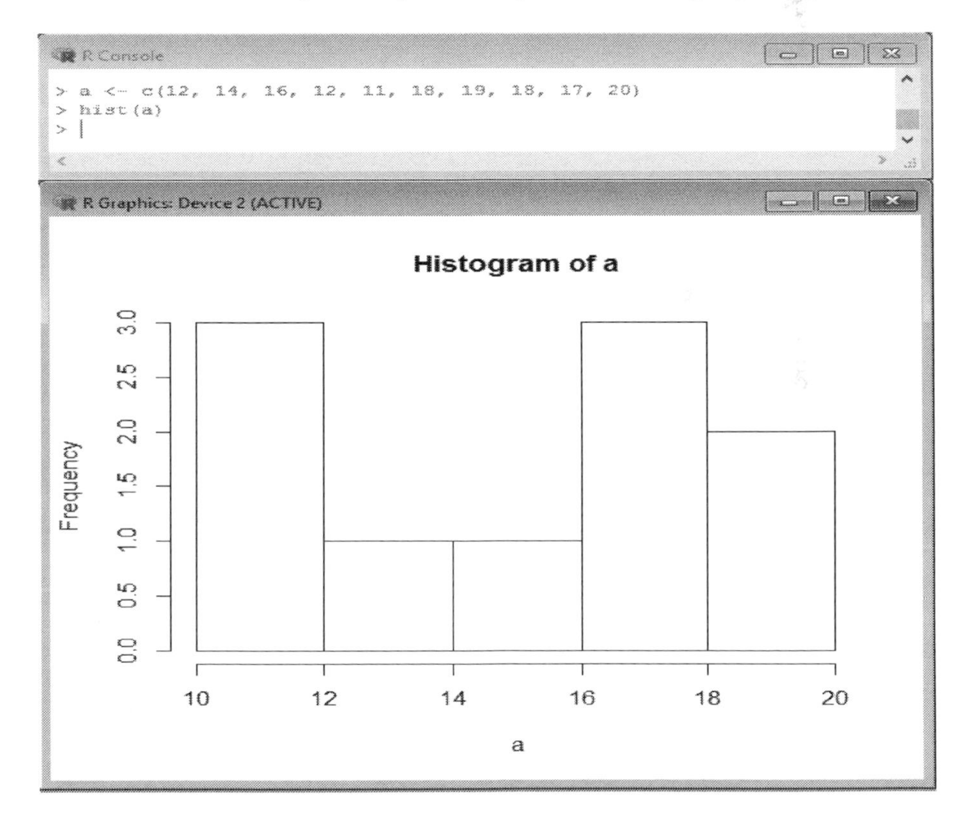

FIGURE 5.18: Illustration of plotting histogram in R

For example, the Figure 5.20, the sort function sorts the data and presents it in lines with the help of 'l' parameter.

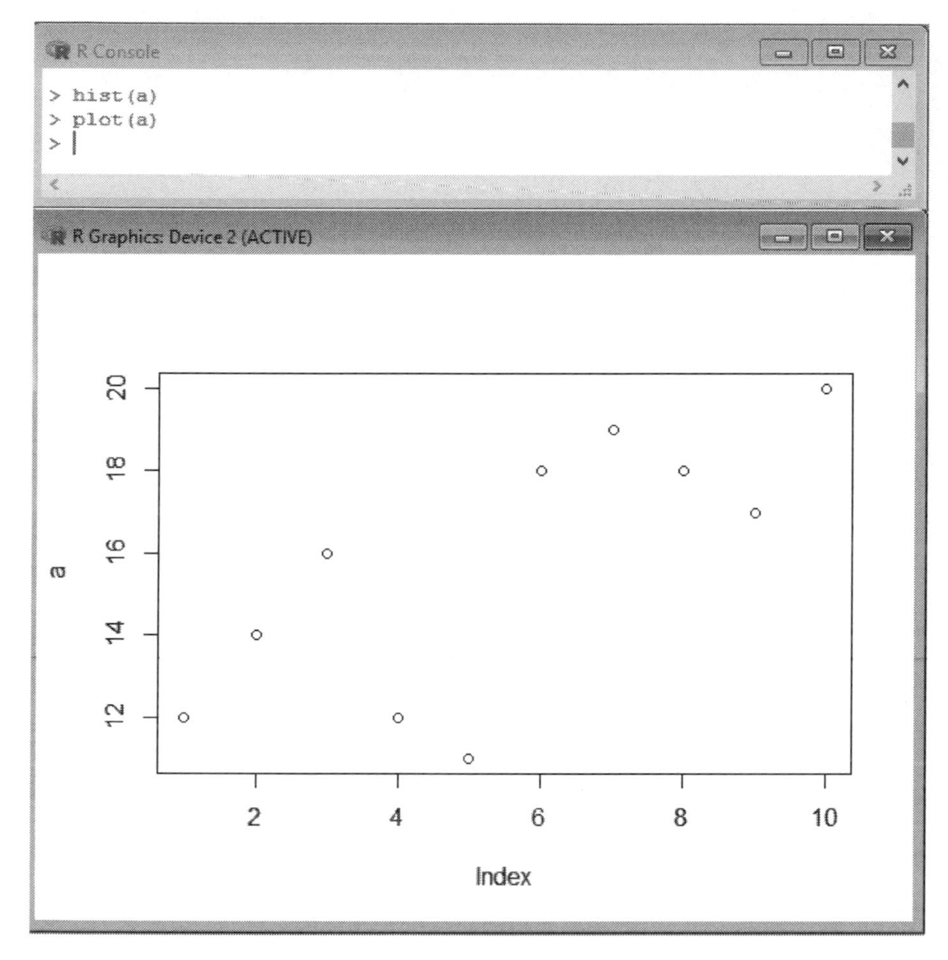

FIGURE 5.19: Illustration of scatter plotting in R

5.3.4.3 Pie Chart

The pie chart is created using the pie() function. The bar or dot plots over pie charts are mostly preferred because vertical judgment is more accurate than the vertical representation while analyzing the data. Pie charts function takes the form, pie(x, labels=), where x is a non-negative numeric vector indicating the area of each slice, and labels= notes a character vector of names for the slices.

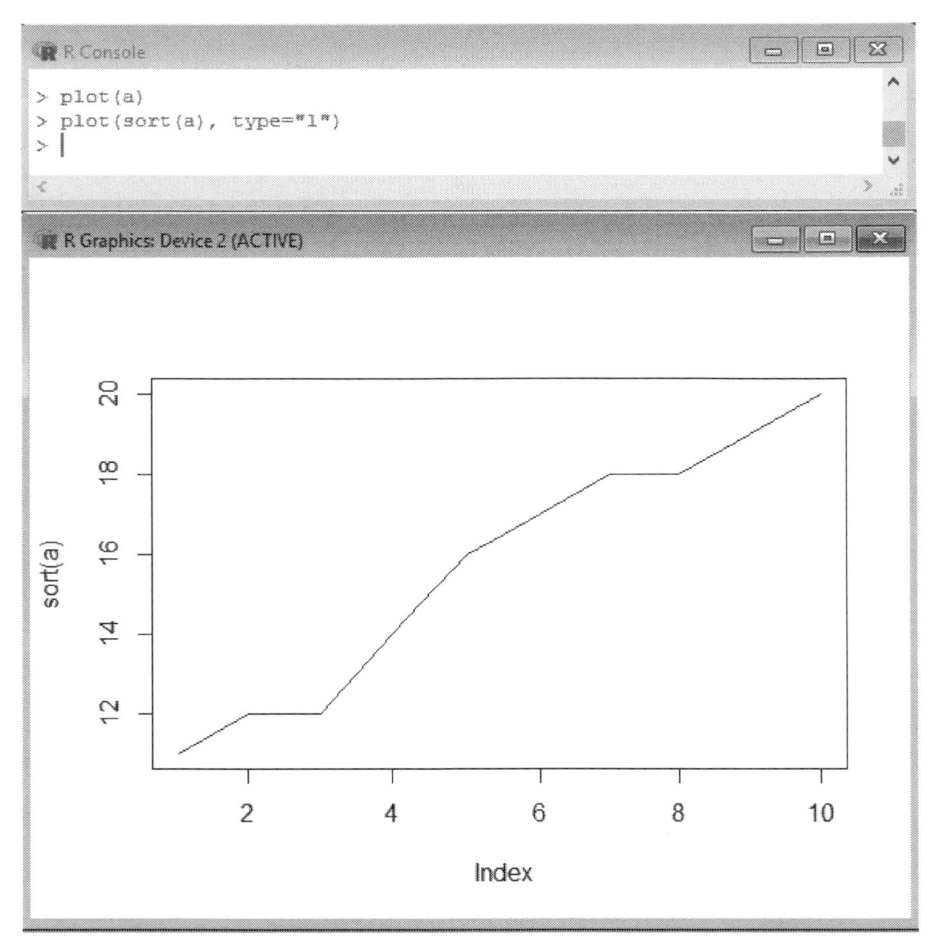

FIGURE 5.20: Illustration of plotting sorted graph in R

5.3.5 Basic Statistics

Once your information has been correctly organized and you have started visually exploring the information, the next stage is typically to define numerically the distribution of each variable, followed by an exploration of the interactions between chosen factors two at a time. There are situations when you want to analyze the series to conclude the result. For example,

- We want to monitor our monthly expenditures. For better analysis, we would like to compute the mean, median, and standard deviation.
- What's the relation between monthly expenses and saving ratio?
- Which group of materials are most costly? And how are they statistically significant?

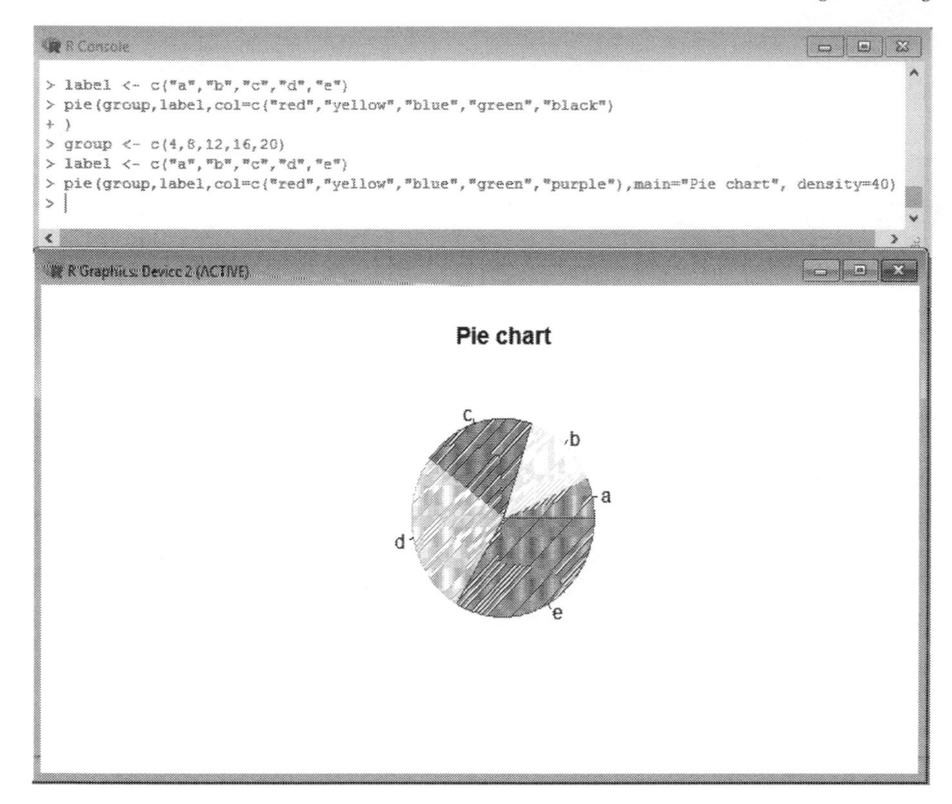

FIGURE 5.21: Illustration of drawing pie chart in R

These questions can be answered by inferring statistics. R has some inbuilt capabilities for doing so.

5.3.5.1 Descriptive Statistics

Descriptive statistics measure central tendency, variability, and distribution shape for continuous variables. The functions (rnorm, sd, var, mean, median, quartile) are easy to perform with the help of a defined name. These functions are elaborated on in Figure 5.22.

5.3.5.2 Correlation Testing

To perform a different statistical test in R, all you need is to download a few packages for the efficient processing of test results. The first package required is:

- fbasics: By installing this package, the dependencies get automatically installed, i.e., 'timeSeries', 'gss', and 'stabledist'.

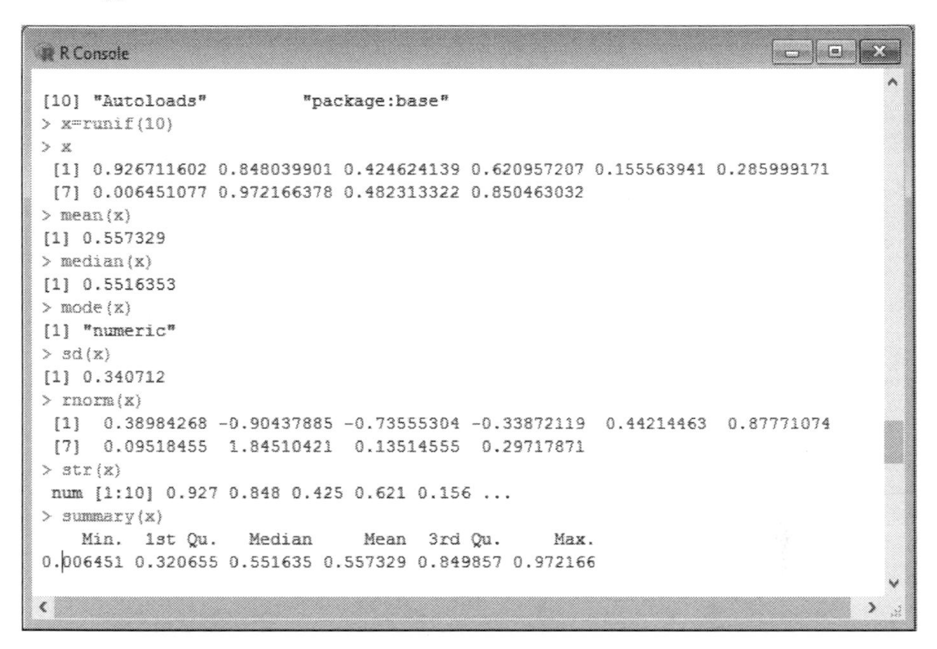

FIGURE 5.22: Examples of descriptive statistics functions

The command used is as shown in Figure 5.23.
>install.packages("fBasics")
Now check the successful installation of the package by the command as shown in Figure 5.24.
>require(fBasics)

- resettest: The resettest package can be installed using the command as shown in Figure 5.25.
  >install.packages("lmtest")
  >require(lmtest)
  This command is used for loading the lmtest package as shown in Figure 5.26.

1. Pearson's correlation test: The functions, cor.teststats, and correlation-Testfbasics can be used to perform this test. Now, all you need is the two correlated sets. First, two vectors are created, then the Pearson correlation command is fired as shown in Figure 5.27.

2. Spearman correlation test: The functions,cor.teststats and spearman.testpspearman can be used to perform this test. The correlated sets are used in the function of Spearman test as shown in Figure 5.28.

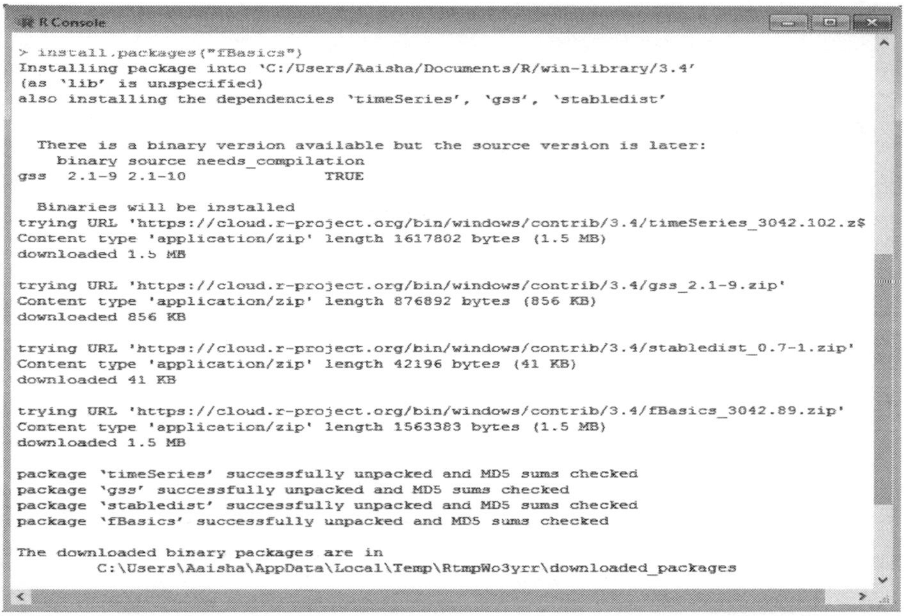

```
R Console
> install.packages("fBasics")
Installing package into 'C:/Users/Aaisha/Documents/R/win-library/3.4'
(as 'lib' is unspecified)
also installing the dependencies 'timeSeries', 'gss', 'stabledist'


  There is a binary version available but the source version is later:
    binary source needs_compilation
gss  2.1-9 2.1-10              TRUE

  Binaries will be installed
trying URL 'https://cloud.r-project.org/bin/windows/contrib/3.4/timeSeries_3042.102.z$
Content type 'application/zip' length 1617802 bytes (1.5 MB)
downloaded 1.5 MB

trying URL 'https://cloud.r-project.org/bin/windows/contrib/3.4/gss_2.1-9.zip'
Content type 'application/zip' length 876892 bytes (856 KB)
downloaded 856 KB

trying URL 'https://cloud.r-project.org/bin/windows/contrib/3.4/stabledist_0.7-1.zip'
Content type 'application/zip' length 42196 bytes (41 KB)
downloaded 41 KB

trying URL 'https://cloud.r-project.org/bin/windows/contrib/3.4/fBasics_3042.89.zip'
Content type 'application/zip' length 1563383 bytes (1.5 MB)
downloaded 1.5 MB

package 'timeSeries' successfully unpacked and MD5 sums checked
package 'gss' successfully unpacked and MD5 sums checked
package 'stabledist' successfully unpacked and MD5 sums checked
package 'fBasics' successfully unpacked and MD5 sums checked

The downloaded binary packages are in
        C:\Users\Aaisha\AppData\Local\Temp\RtmpWo3yrr\downloaded_packages
```

FIGURE 5.23: Steps to follow for fBasics package installation

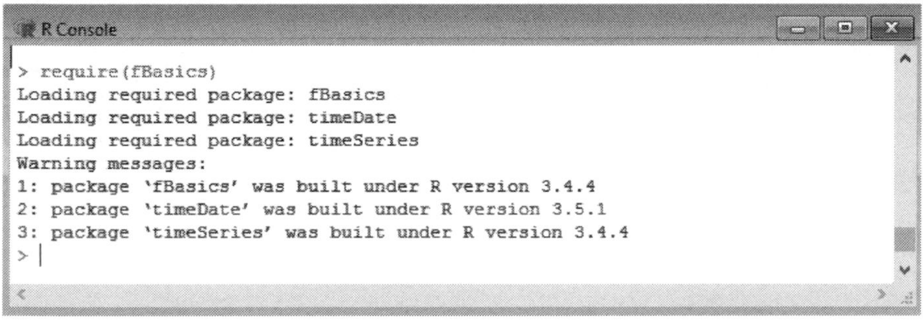

```
R Console
> require(fBasics)
Loading required package: fBasics
Loading required package: timeDate
Loading required package: timeSeries
Warning messages:
1: package 'fBasics' was built under R version 3.4.4
2: package 'timeDate' was built under R version 3.5.1
3: package 'timeSeries' was built under R version 3.4.4
>
```

FIGURE 5.24: Steps to follow for loading fBasics package

3. Smirnov test: This test is used to check that if two samples have been generated from the same probability distribution as shown in Figure 5.29. The function ks.teststatscan is used to perform this test.

4. Chi-squared test: The package 'outliers' is required for the execution of this test. This test can be performed by using the function chisq.out.testoutliers with the form chisq.out.test(data, variance=1). The population variance is named as parameter variance in this test. This test is performed on two variable as discussed in Figure 5.30.

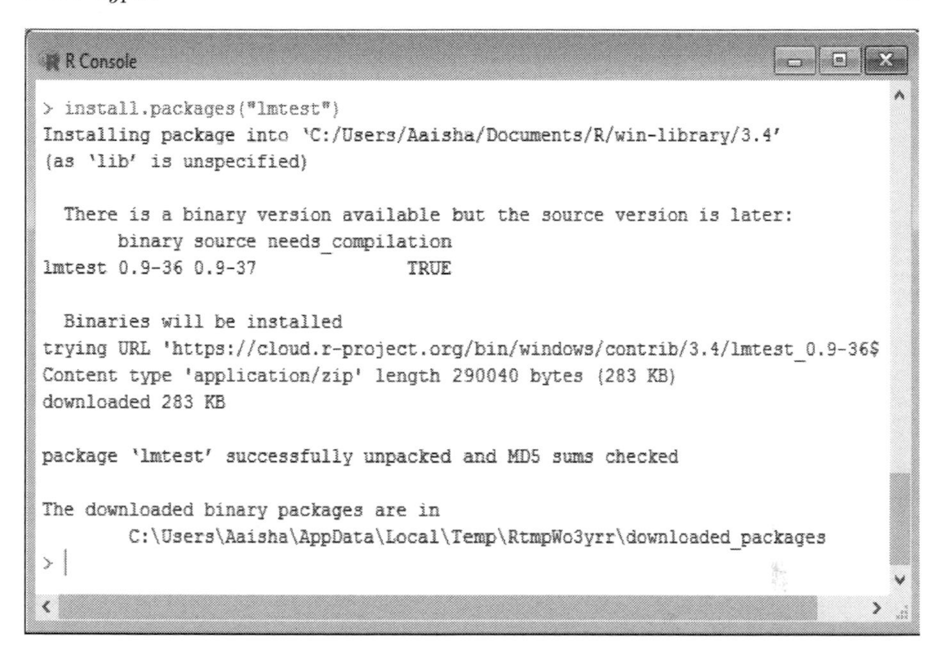

FIGURE 5.25: Steps to follow for installing lmtest package

```
R Console
> a <- c(12.5, 14, 16, 12, 11, 18, 19, 18.5, 17, 20)
> b <- c(2, 3.5, 7, 5.5, 4.5, 8, 9.5, 8.5, 4, 6)
> cor.test(a,b,method="pearson",alternative="two.sided",conf.level = 0.95)

        Pearson's product-moment correlation

data:  a and b
t = 2.5849, df = 8, p-value = 0.03237
alternative hypothesis: true correlation is not equal to 0
95 percent confidence interval:
 0.07820431 0.91541363
sample estimates:
     cor
0.674613

> correlationTest(a, b)

Title:
 Pearson's Correlation Test

Test Results:
  PARAMETER:
    Degrees of Freedom: 8
  SAMPLE ESTIMATES:
    Correlation: 0.6746
  STATISTIC:
    t: 2.5849
  P VALUE:
    Alternative Two-Sided: 0.03237
    Alternative      Less: 0.9838
    Alternative   Greater: 0.01618
  CONFIDENCE INTERVAL:
    Two-Sided: 0.0782, 0.9154
         Less: -1, 0.8939
      Greater: 0.1949, 1
```

FIGURE 5.26: Steps to follow for loading lmtest package

```
R Console
> a <- c(12.5, 14, 16, 12, 11, 18, 19, 18.5, 17, 20)
> b <- c(2, 3.5, 7, 5.5, 4.5, 8, 9.5, 8.5, 4, 6)
> cor.test(a,b,method="pearson",alternative="two.sided",conf.level = 0.95)

        Pearson's product-moment correlation

data:  a and b
t = 2.5849, df = 8, p-value = 0.03237
alternative hypothesis: true correlation is not equal to 0
95 percent confidence interval:
 0.07820431 0.91541363
sample estimates:
     cor
0.674613

> correlationTest(a, b)

Title:
 Pearson's Correlation Test

Test Results:
  PARAMETER:
    Degrees of Freedom: 8
  SAMPLE ESTIMATES:
    Correlation: 0.6746
  STATISTIC:
    t: 2.5849
  P VALUE:
    Alternative Two-Sided: 0.03237
    Alternative      Less: 0.9838
    Alternative   Greater: 0.01618
  CONFIDENCE INTERVAL:
    Two-Sided: 0.0782, 0.9154
         Less: -1, 0.8939
      Greater: 0.1949, 1
```

FIGURE 5.27: Description of the procedure to perform Pearson correlation test

5.3.6 Packages

The R comes out of the box with comprehensive capacities. But, some of its most interesting characteristics can be downloaded and installed as optional modules. You can download from http:/cran.r-project.org/web/packages more than 5,500 user-contributed modules called packages. Packages are the collection of libraries, which further contain defined functions. They provide a tremendous variety of fresh capacities, from geospatial data assessment to the processing of protein mass spectra to psychological testing assessment! You can manipulate packages with a number of R features. Use the install.packages command to first install a package. For instance, install.packages, provides a list of CRAN mirror locations without choices. Once you select a site, a list of

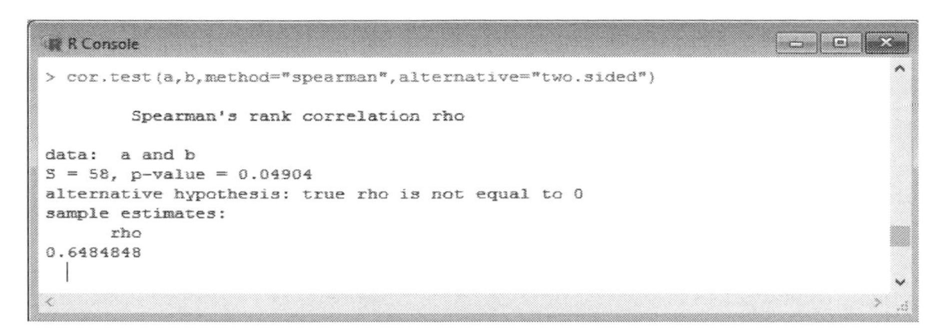

FIGURE 5.28: Description of the procedure to perform Spearman correlation test

all accessible packages will be provided to you. Select and install one download. Following are the commands which are used with Packages in R. These commands are experimented and presented in Figure 5.31.

- install.packages(): This command is to install packages in the library. In the figure, we have installed "vcd" package.
- library(): It is downloaded from a CRAN mirror site and placed in your library by installing a package. You must load the package using the library. It is done with the library() function. We have loaded the same installed package.
- search(): It is exciting to view the existing installed packages in R with a single command, i.e., search().

5.3.7 Input Parameters Formats for R

Till now, we have learned to create different data structures. R has the capability to let the user enter the data from different sources of different formats. This feature helps to analyze a large amount of data and draw the results for numerous experiments and dedicated prediction. For this, R supports a variety of tools. The manual is available at http://mng.bz/urwn. A few important methods for fetching the input are discussed below:

1. Runtime: Entering the data during the execution of the program is one of the historical methods. R supports the text editor, which is invoked with the help of the edit() function. Following are the steps to be used for entering the data with this editor:

   (a) Create the data structure which holds the required variable name. This step would create only the structure of data types such as data frame, array, or matrix.

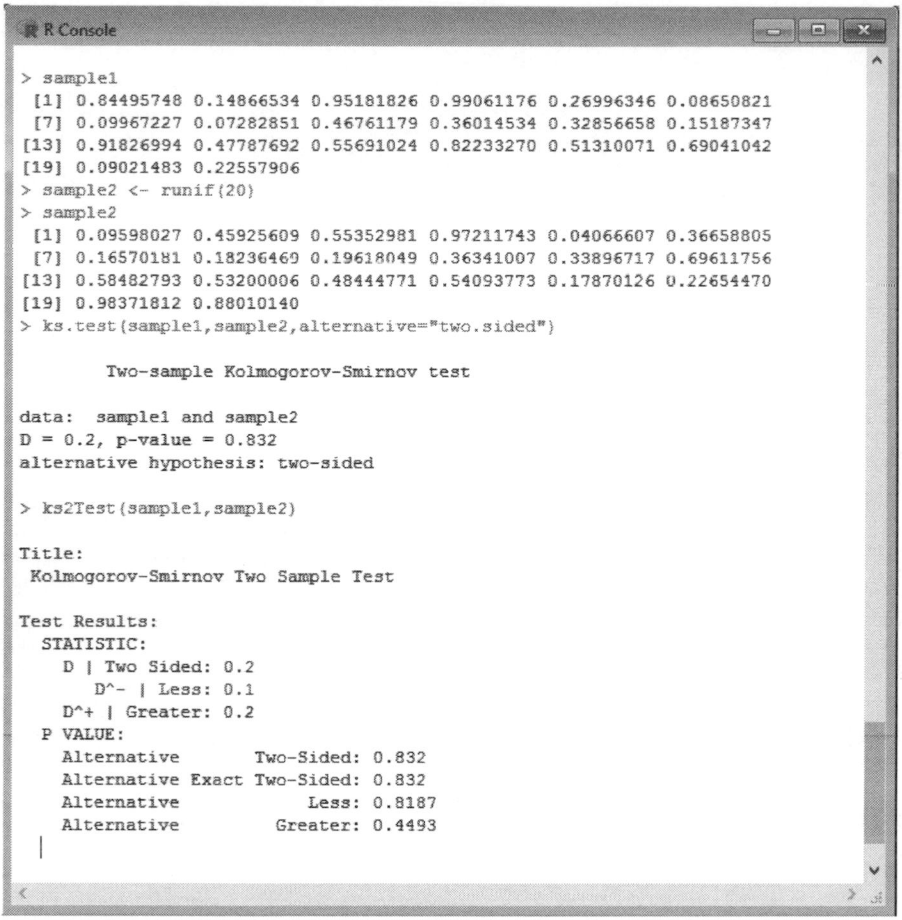

```
R Console
> sample1
 [1] 0.84495748 0.14866534 0.95181826 0.99061176 0.26996346 0.08650821
 [7] 0.09967227 0.07282851 0.46761179 0.36014534 0.32856658 0.15187347
[13] 0.91826994 0.47787692 0.55691024 0.82233270 0.51310071 0.69041042
[19] 0.09021483 0.22557906
> sample2 <- runif(20)
> sample2
 [1] 0.09598027 0.45925609 0.55352981 0.97211743 0.04066607 0.36658805
 [7] 0.16570181 0.18236469 0.19618049 0.36341007 0.33896717 0.69611756
[13] 0.58482793 0.53200006 0.48444771 0.54093773 0.17870126 0.22654470
[19] 0.98371812 0.88010140
> ks.test(sample1,sample2,alternative="two.sided")

        Two-sample Kolmogorov-Smirnov test

data:  sample1 and sample2
D = 0.2, p-value = 0.832
alternative hypothesis: two-sided

> ks2Test(sample1,sample2)

Title:
 Kolmogorov-Smirnov Two Sample Test

Test Results:
  STATISTIC:
    D | Two Sided: 0.2
       D^- | Less: 0.1
    D^+ | Greater: 0.2
  P VALUE:
    Alternative       Two-Sided: 0.832
    Alternative Exact Two-Sided: 0.832
    Alternative            Less: 0.8187
    Alternative         Greater: 0.4493
```

FIGURE 5.29: Description of the procedure to perform smirnov test

(b) Enable the editor with the edit command and enter the data.

Both of these steps are illustrated in Figure 5.32. In this figure, the first screen is used for creating the empty data frame and, invoking the editor with this frame, popping up the second screen.

Now, you can easily enter the data in the cells. Once you close the data table window, it is automatically saved. There is no limitation on the number of times editing a particular structure. You can do this again with the help of edit(). Even the fix() the function works similarly as shown in Figure 5.33.

2. Text file: Importing the data from the text file can be done with the help of the read.table() function. It reads the file in the format of the table and saves it in the form of a data frame. One line of the file is equivalent to one

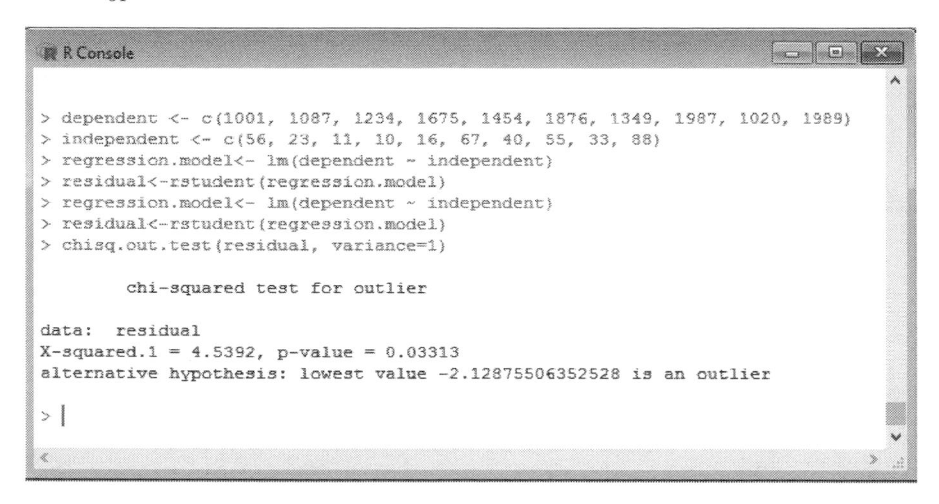

FIGURE 5.30: Description of the procedure to perform chi-squared test

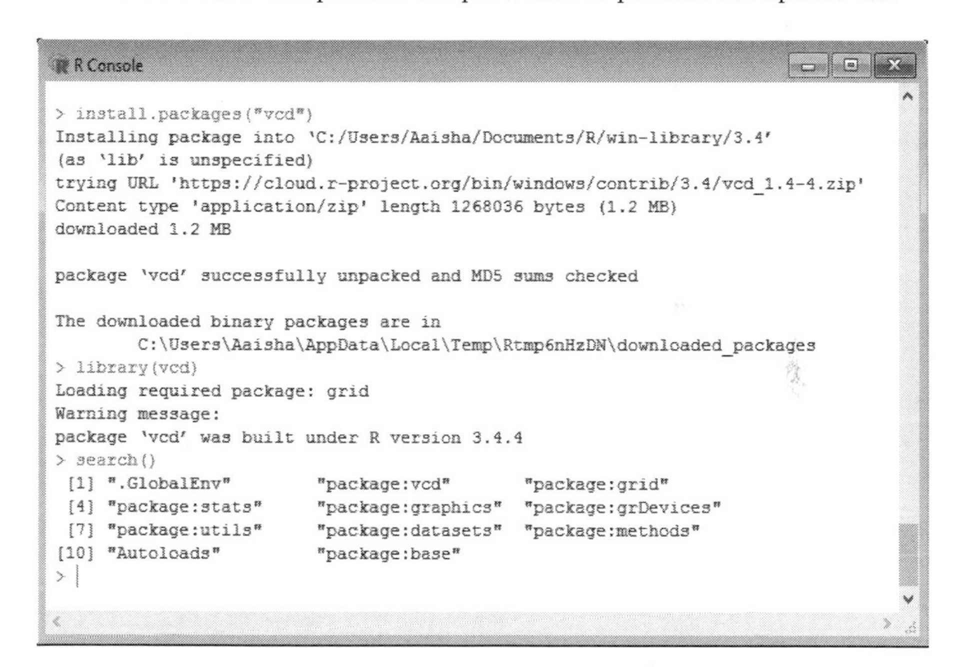

FIGURE 5.31: Representation of functions to view, search, and install packages in R

row of the table. The syntax of importing the data using this command is:
newdataframe<- read.table(file, options)
There is a list of options which can be used in Table 5.4.

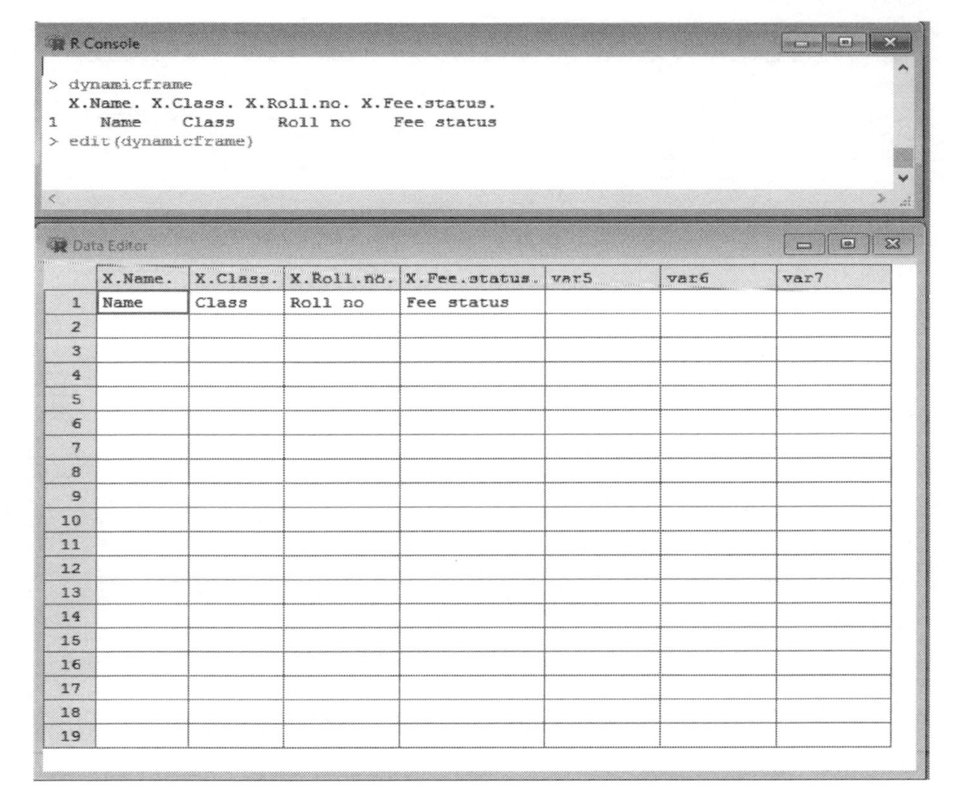

FIGURE 5.32: Procedure to follow for getting the input data during runtime

TABLE 5.4: Options of read.table function

Option	Description
quote	Character which is used to delimit the strings that contain special characters. By default the value of this is quote is either double (") or single (') quotes.
header	It is logical, which indicates the occurrence of a variable name in the header of the file, i.e., the first line.
sep	It is used for reading the file, which may contain any space tool, such as tab, carriage return, newline, and comma. The comma is most preferred as the maximum of the file are comma deliminators. In the case of a comma, it is sep= ","
row.names	An optional parameter specifying one or more variables to represent row identifiers.
col.names	If the first row of the data file doesn't contain variable names (header=FALSE), you can use col.names to specify a character vector containing the variable names. If header=FALSE and the col.names option is omitted, variables will be named V1, V2, and so on.

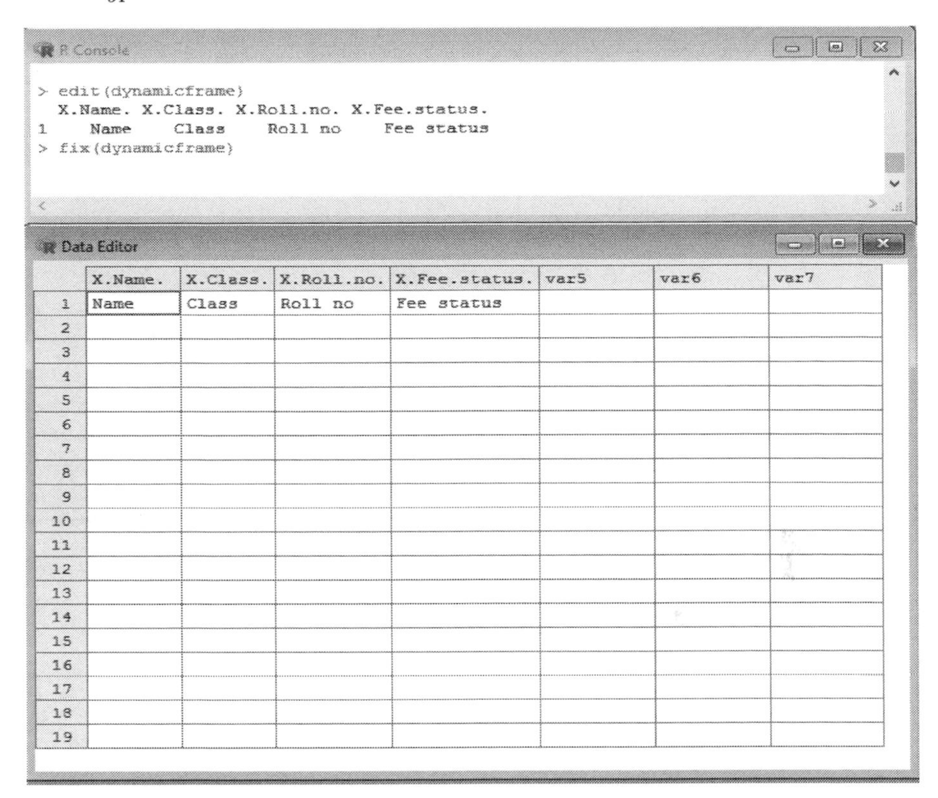

FIGURE 5.33: Procedure to follow for getting the input data during runtime with fix function

3. Excel file: It is simple to import the data from an Excel file by the R package named "xlsx". This package helps to read, write, and format Excel 97/2000/XP/ 2003/2007 files. A data frame is created by importing the data with read.xlsx() function. The syntax of this function is as follows.
   read.xlsx(file, n), where the file is the complete Excel file to be imported and n is the count of the files to be imported. You need to install two more packages named xlsxjars and rJava. For example:
   install.packages("xlsx")
   library(xlsx)
   workbook <- "C:/result.xlsx"
   newdataframe<- read.xlsx(workbook, 1)
   These commands perform the following action:

   (a) The package named "xlsx", is installed.
   (b) The package gets loaded in the memory.
   (c) The path of the Excel file is determined and saved in another variable named workbook.

(d) The data is read from the workbook and loaded in the data frame named newdataframe.

We can define the specific number of rows and columns to be imported with the parameters rowIndex and colIndex to the read.xlsx function.

4. Database management system: R can interface with a broad range of DBMS related database management systems, including Microsoft Access, MySQL, Microsoft SQL Server, Oracle, DB2, Teradata, Sybase, and SQLite. Some packages provide access via indigenous database drivers, while others provide access via ODBC or JDBC. The RODBC package, which enables R to connect to any DBMS that has an ODBC driver, may be the most common technique of accessing a DBMS in R. This involves all the previously mentioned DBMSs. The first phase is to install and configure the suitable ODBC driver (these drivers are not components of R) for your platform and database. If the required drivers are not already installed on your machine, an Internet search can provide you with options. Install the RODBC package once the drivers are installed and configured for the database(s) of your choice, using the install.packages("RODBC) "command. There is a list of functions supported by RODBC as discussed below:

   - odbcConnect(dsn,uid="",pwd="") : It is used for establishing a connection to an ODBC database.
   - sqlFetch(channel,sqltable): A data frame is created by reading a table from an ODBC database.
   - sqlQuery(channel,query): It is used for firing a query to the database, and the returns are presented.
   - sqlSave(channel,mydf,tablename= sqltable,append=FALSE): It is used for updating the ODBC database by writing or updating a data frame.
   - sqlDrop(channel,sqltable): This command is used for removing a table from the database.
   - close(channel): Finally the connection is closed.

5.4 Summary and What's Next?

The requirement of learning R programming language is not only essential for R but also for individual programming. This chapter is sufficient for the learner to clear the basic concepts of R. The basic commands of R are well illustrated with examples. The data types of R are elaborated with programs. The operators are also well defined. The role of graphics in R programming is explained with the help of different plots. The basic statistical functions are described along with their usage. The shell of R programming, i.e., packages, are also covered in this chapter. At the end of this chapter, the learner would be well familiar with R programming language.

5.5 Exercises

Questions

1. Define R. What are the features of R programming?
2. How we can assign the values to different variables?
3. What are the various ways of adding comments to the R program?
4. Elaborate on the different data types used in R.
5. What are matrices? How is the element inserted and retrieved in a matrix?
6. What are the different operators used in R?
7. Define descriptive statistics and their usage in R programming.
8. What are packages and how do you install them?
9. What are the different ways of getting input from the user for the R program?
10. Define vectors and their usage in R.

Multiple Choice Questions

1: Which symbol is used for assigning a value to the variable?

(a) <- **(b)** <=

(c) = **(d)** ==

2: Which symbol is used for adding comments to the R program?

(a) % **(b)** #

(c) * **(d)** $

3: Which value is accepted by the variable of number data type?

(a) Integer **(b)** Real value

(c) Decimal value **(d)** All of the above

4: What is the role of dimensions in the array() function?

(a) List of value **(b)** Index of the array

(c) Dimension list **(d)** None of them

5: What kind of data can a data frame can hold?

(a) String **(b)** Integer

(c) Vector **(d)** All of them

6: Which operator is used for the multiplication of matrices?

(a) * **(b)** mul()

(c) %*% **(d)** None of them

7: Which operator returns the remainder?

(a) It is used for ordering the different data structure. **(b)** It is used for ordering the data.

(c) It is used for ordinal variables. **(d)** None of these

8: A vector is a one-dimensional array which holds the first value at the index:

(a) Zero **(b)** One

(c) Minus one **(d)** None of them

9: What kind of data can vector hold?

(a) Numeric **(b)** Character

(c) String **(d)** All of above

10: What is the role of "==" operator?

(a) Assignment **(b)** Comparison

(c) Bitwise shift **(d)** None of the above

11: Which operator is used for performing a logically AND operator?

(a) & **(b)** |

(c) && **(d)** ||

12: runif() function is used for?

(a) Running the program continuously **(b)** Running the program until a particular condition is met

(c) Generates some random numbers with a specified limit **(d)** None of the above

13: Which function is used for loading the package?

(a) install.packages() **(b)** packages()

(c) library() **(d)** None of the above

14: Which function is used to check the list of install.packages?

(a) install.packages() **(b)** search()

(c) find() **(d)** None of the above

15: Which command is used for opening the editor with the purpose of entering the input?

(a) enter() **(b)** edit()

(c) file() **(d)** retrieve()

16: Which function is used for entering the data in the table format via a text editor?

(a) edit.table() **(b)** read.table()

(c) write.table() **(d)** table()

17: Which package is required for fetching the data from the Excel file?

(a) xlsx **(b)** xlsxjars

(c) rJava **(d)** All of the above

18: Which function is used for reading the data from the Excel file?

(a) read.xlsx()

(b) get.xlsx()

(c) fetch()

(d) None of the above

19: Which function is used for reading the data from the sql file?

(a) sqlRead()

(b) sqlGet()

(c) sqlFetch()

(d) None of the above

20: Which command is used for closing the connection established with the database?

(a) exit()

(b) close()

(c) terminate()

(d) None of the above

6

Machine Learning Paradigms

6.1 Introduction

Just imagine, the complete surroundings in the form of data. We can say that we are engulfed with data. Without any doubt, this data will keep on increasing in the near future. Now, who all are facing problems with data? The actual problem is for who needs it? Firstly the Industrialists and producers, who want to record the present sale and consumptions. Secondly, the research team, who want to predict the demand of the products in future. It is impossible to store such volume of data on a personal desktop. Let's take the example, you go to the supermarket and purchase a few household items. There can be multiple counters for billing. So, you can proceed for billing and analyze in whatever counter you reach, the billing can be done. So, all the items can be recorded on the server and fetched using bar codez. It is quite obvious that if the volume of data increases, the computing complexity also increases. The hidden insights of the data thus remain unexplored. Here, the role of machine learning comes into action. A machine learning scientist aims is to gather the meaningful data by discovering the hidden patterns that rule this world and bind them in a procedure which helps in predicting the scenario in future new conditions.

Structural description. As discussed in the previous chapter, the IoT dataset is used throughout this book. So we are referring to the same dataset. How can I describe the dataset in the structure? Does it make any sense? We use a snapshot of data in Figure 6.1 for better understanding.

If wind direction increases, then the speed of
wind speed decreases and the chance of rain increases

It is not necessary to design the structural description like this. Such a rule which requires decisions according to different situations can be drawn by a decision tree. A decision tree is a scientific algorithm which states the sequence of different decisions by various conditions. The example we have considered is the simple one by just referring to 25 rows, but actually, this dataset has 59145 rows. The structural description drawn does not generalize from the complete data. There are three reasons behind this incorrect description.

	Wet Bulb Temperature	Humidity	Rain Intensity	Interval Rain	Total Rain	Precipitation Type	Wind Direction	Wind Speed	Maximum Wind Speed	Solar Radiation	Heading	Battery Life	Measu
2	7	55	0	0	1.4	0	63	1.9	2.8	780	322	12	######
3	6.3	56	0	0	1.4	0	124	1.5	2.3	180	322	12.1	######
4	6.5	54	0	0	1.4	0	156	1.9	3.4	127	322	12.1	######
5	6.3	53	0	0	1.4	0	150	1.4	4.5	67	322	12.1	######
6	6.4	52	0	0	1.4	0	155	1.1	2.3	10	322	12	######
7	7.4	58	0	0	1.4	0	175	1.3	2.5	2	322	12	######
8	7.5	58	0	0	1.4	0	164	1.7	3.5	3	322	12	######
9	7.1	45	0	0	1.4	0	155	0.7	2.3	2	322	12.1	######
10	9	32	0	0	1.4	0	159	2	3.5	434	322	12.1	######
11	10.9	42	0	0	1.4	0	155	2.3	3.9	799	322	12.1	######
12	11.6	62	0	0	1.4	0	85	2	2.8	850	322	12.1	######
13	11.2	55	0	0	1.4	0	109	2.5	3.8	900	322	12.1	######
14	12.4	43	0	0	1.4	0	147	2	3	53	322	12.1	######
15	11.7	50	0	0	1.4	0	137	2.2	3.3	887	322	12	######
16	12.5	45	0	0	1.4	0	136	2	3.5	563	322	12.1	######
17	12.4	40	0	0	1.4	0	150	1.1	3	44	322	12.1	######
18	13.4	33	0	0	1.4	0	7	1	1.4	0	322	12.1	######
19	13.7	55	0	0.1	1.5	60	193	0.4	0.8	43	322	12.1	######
20	14	68	0	1	2.5	0	119	0.7	1.1	59	322	12.1	######
21	14	76	0	0.1	2.5	0	345	0.6	1.3	120	322	12	######
22	13.4	63	0	0	2.6	0	145	2.3	3.5	358	322	12.1	######
23	12.6	70	0	0	2.6	0	137	1.1	3	266	322	12.1	######
24	14.3	73	1.2	0.4	3	60	147	1	3.8	40	322	12	######
25	15.7	82	0	0	11.1	0	148	2.4	4.3	1	322	12.1	######

FIGURE 6.1: Findings of structural description

1. The complete data has not been taken into consideration while making this decision.
2. Only the three features, wind direction, wind speed, and rain have been considered.
3. There may be some noise or missing data in the rest of the rows.

The machine learning models work with the input, on the basis of the relation among the features and predicts the output. The input has many forms, such as attributes and instances. How the input is configured is commonly known as the concept description. The conceptual data or input is used for exchanging in this book. It is the first thing a machine learns for the next stage of modeling. It is the part of learning that the system adapts. Providing the philosophy of the internal working of the machine would be wrong to pin down in this book, and all the working we are concerned with is conceptual and logical. It means we are only trying to figure out how the machine works, what kind of input it takes, and what kind of output it produces. Yes, knowing the learning process is important because it is the only way to judge the machine's intelligence. In the next section, we will discuss different inputs, the processed output, and the various paradigms of learning.

6.2 Generalizing Input

The instance is an individual, an independent, and a fixed value which is characteristic by the various features or attributes. Each attribute is a feature which can measure the instances or we can call them rows. Every problem has definite features. Let's talk about weather forecasting. Just sitting at home, and browsing your cell phone, you get to know the weather of next month. How such prediction became possible, it is due to the hard work of machine learning scientists. Let us consider the weather data as shown in Table 6.1. The input to the machine can be classified as clustered or associated. We can say how the atmosphere is by evaluating using attributes of temperature, wind, and humidity. The models are experimented with the matrix of instances constituting the attributes as columns. Thus, the dataset is fed into the simple .csv file, popularly called a flat-file.

TABLE 6.1: Weather forecasting using numerical data

Outlook	**Temperature**	**Atmosphere**	**Wind**	**Humidity**	**Day**
Sunny	87	Hot	No	90	Sunday
Cloudy	67	Cold	Yes	80	Monday
Rainy	74	Cold	No	65	Tuesday
Moderate	80	Normal	Yes	75	Wednesday
Moderate	82	Normal	No	90	Thursday
Sunny	94	Hot	No	90	Friday
Cloudy	72	Cold	No	90	Saturday

The data can be expressed as trees also because of the relationships among them. For example, the data of family including ancestors can be drawn in the form of a tree, and this tree would be treated as an input for the learning of machine. I mean by far relatives such as second cousin maternal relatives. To solve this problem, consider Figure 6.2 in which two categories can be formed. First is to find the blood relations, second is to find the other relations, and third is to find no relation.

1. Blood relations:

Me —-> Father
Me —-> Mother
Father —-> Father's father
Father —-> Father's mother
Mother —-> Mother's mother
Mother —-> Mother's father
Father —-> Father's Paternal Grandmother
Father —-> Father's Paternal Grandfather

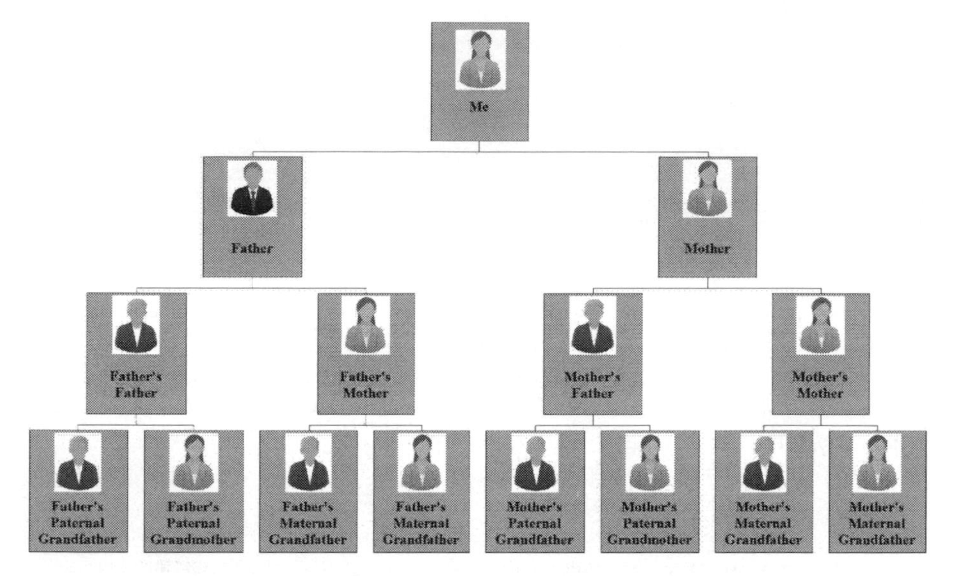

FIGURE 6.2: Complete family tree in the hierarchical format

Mother —-> Mother's Paternal Grandmother
Mother —-> Mother's Paternal Grandfather

2. Other relations:

Father —-> Mother
Father's father —-> Mother's father
Father's father —-> Mother's mother
Father's mother —-> Mother's father
Father's mother —-> Mother's mother
Father —-> Father's Maternal Grandmother
Father —-> Father's Maternal Grandfather
Mother —-> Father's Paternal Grandmother
Mother —-> Father's Paternal Grandfather
Mother —-> Father's Maternal Grandmother
Mother —-> Father's Maternal Grandfather
Mother —-> Mother's Maternal Grandmother
Mother —-> Mother's Maternal Grandfather

3. No relation: As it is the family tree, so every family member is related to each other in one or the other way.

Now, to make this example more understandable, we need to convert this tree into the table. Again, we can notice the instances are somehow related to each other. Let's begin with constructing Table 6.2. This example shows how we can take a relationship between different nodes of a tree and recast it into a set of independent instances. Every instance has a fixed define value, characterized by attributes. This is how the instances and attributes form inputs for the machine learning model.

6.3 Generalizing Output

The simplest way to represent output after machine learning is the same as the input. The decision table represents the attributes with instances as input and produces output in the same manner. The decision Table 6.2 representing the family tree also represents output in the sense that it depicts the relationships among the instances and attributes. Not all the attributes are considered; the most trivial condition is to select which attributes should be considered.

6.3.1 Decision Tree

A decision tree is a structure that represents the set of independent instances by extracting the relation between them. This follows a popular approach known as 'divide-and-conquer' for learning from these unknown instances to form a decision tree. The previous example of a family is the decision tree. The testing is mainly used by selecting the most vital attribute, such as the discussed example of family tree, by selecting the attribute 'Me' as the parent node. Like the relation of all the family members is compared with the parent node to form the leaf nodes. This selection requires a comparison between two or more nodes to form a basis. The condition we have learned in all programming languages, i.e., if() or else(), that could also be applied while forming the classification rule. This condition is to be applied at the parent node and consecutively to the further child node forming a parent node. Because of the structure of a tree, there are always two decisions to be taken; this is the reason for embedding this condition. Thus, it forms a classification rule. The classification itself determines the instance between two values; same is the case here. The classification rules are explored more in the next section.

In case of the decision which results in two classes, the greater than or less than approach could be followed. If the case is of three branches to be formed, then the three conditions may be tested, i.e., less than, equal to, and greater than. This three-way split may be followed if there are more than two possibilities. If there is a missing value then it may also follow the alternative to create a decision. The case is different for real-valued attributes, in which the equal operation may not be successful. So, the solution for it is the interval

TABLE 6.2: Instances forming a family tree table structure

Relation	a	b	c	d	e	f	g	h	i	j	k	l	m	n	o
Me (a)	√	√	√	√	√	×	×	×	×	×	×	×	×	×	×
Father (b)		√	√	√	×	×	×	×	×	×	×	×	×	×	×
Mother (c)	√	×	×	√	√	×	×	×	×	×	×	×	×	×	×
Father's father (d)	×	×	×	√	√	√	×	×	×	×	×	×	×	×	×
Father's mother (e)	×	×	×	√	√	×	×	×	×	×	×	×	×	×	×
Mother's father (f)	√	√	√	√	√	√	×	×	×	×	×	×	×	×	×
Mother's mother (g)	√	√	√	√	√	√	×	×	×	×	×	×	×	×	×
Father's paternal grandfather (h)	√	√	√	√	√	×	×	×	×	×	×	×	×	×	×
Father's paternal grandmother (i)	√	√	√	√	√	×	×	×	×	×	×	×	×	×	×
Father's maternal grandfather (j)	√	√	√	√	√	×	×	×	×	×	×	×	×	×	×
Father's maternal grandmother (k)	√	√	√	√	√	×	×	×	×	×	×	×	×	×	×
Mother's paternal grandfather (l)	√	√	√	√	√	×	×	×	×	×	×	×	×	×	×
Mother's paternal grandmother (m)	√	√	√	√	√	×	×	×	×	×	×	×	×	×	×
Mother's maternal grandfather (n)	√	√	√	√	√	×	×	×	×	×	×	×	×	×	×
Mother's maternal grandmother (o)	√	√	√	√	√	×	×	×	×	×	×	×	×	×	×

class. The three alternatives for the internal class is below, above, and within. This is how we conclude the decisions from the structure of tree.

6.4 Classification Rules

Classification rules are the decision to be taken between two alternatives. The pre-condition or the post-condition are the tests to be performed simultaneously. Just like the decision tree, we can say this case is somehow executed in the same manner as a probability distribution. When there are multiple preconditions, these conditions are combined using the AND gate. After the complication, the logical rule can be formed, thus forming a creative condition. It is not the compulsion to form a combine multiple conditions using AND. However, there may be conjections in the conditions, so the OR gate can also be implemented. The OR gate results in true when any of the condition is true. If the same procedure is followed, then the conditions become complex and will increase the computational complexity. It is relevant to form the rules with respect to the decision tree criteria. The conditions drawn by the decision trees targets to remove the redundant tests. The rules in the decision tree are not finalized in a simple way because the disjunction applied while forming the conditions is not straightforward. The reason behind these disjunction is because of the similar conditions for all the different attributes.

As observed in the example illustrated in Figure 6.3, the tree is a decision tree. This tree shows that the decisions have been taken while traversing through the instances. The attributes are not considered so deeply because the decisions could be taken by just analyzing the attributes. Such a problem as that occurred in this tree is exclusively known as the *replicated subtree* problem. This issue should be resolved with the help of a better compilation of repeated conditions. The input is the desirable file with the collection of multiple attributes. The output is also the collection of instances but deeply analyzes the structure and patterns of the data. The findings of this example are drawn below.

Figure 6.4, predicts the new rules formed by analyzing. Each rule has some knowledge, which is not common. You can see these rules have been extracted from the given tree. The tree which formed at random actually doesn't follow the rules. New rules can be added to change the structure of a tree. The existing structure is revised by adding a few new rules. The revised tree structure is presented in Figure 6.5. The existing ruleset remains the same, by providing space to the new one.

The tree structure has been revised by adding a few rules; it is not always necessary to revise the rules. Sometimes, the problem arises when an instance is not capable of belonging to any class. This may not happen with decision trees. The reason is that the rules are formed in a way that each instance is

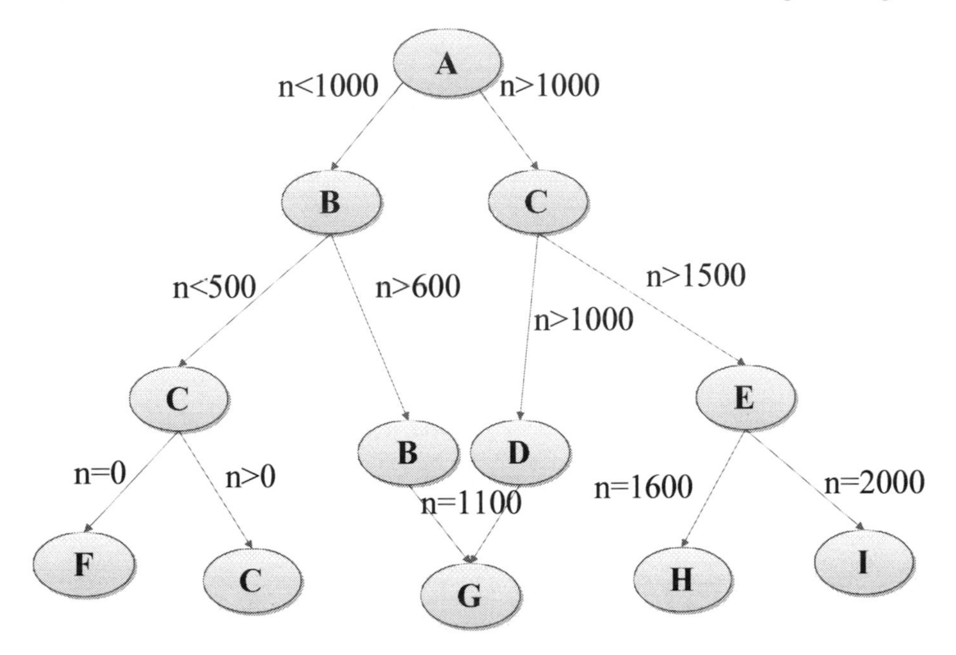

FIGURE 6.3: Randomly decision tree formed using classification rules

Tree observations	
B	600-1000
C	0-500
D	1000-1500
E	Above 1500
F	n=0
G	n=1100
H	n=1600
I	n=2000

FIGURE 6.4: After analyzing the problem of replication

classified in one or the other class. If this happens, there are two solutions for it. The first is to make that instance the parent node; second is to add a new rule which classifies that rule properly. The second one is adopted in our example; the new decision list is presented in Figure 6.6.

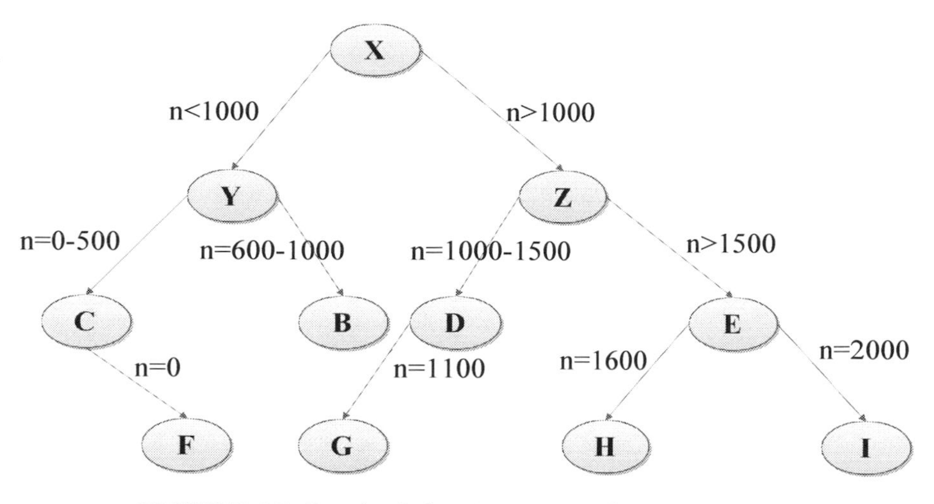

FIGURE 6.5: Resolved decision tree without replication

Revised Rules	
Y	n<1000
Z	n>1000
B	n= 600-1000
C	n= 0-500
D	n= 1000-1500
E	Above 1500
F	n=0
G	n=1100
H	n=1600
I	n=2000

FIGURE 6.6: Decision list of the resolved decision tree

6.5 Numeric Prediction

We have learned how to categorize the instances by drawing new classes. The results predicted the categories so we talk about how to predict the numeric values. Decision trees are used for classification problems. However, to make the machine learn in order to predict the numeric values is similar to the case of regression. Regression is the supervised learning used for predicting real point values. The regression trees can also be used for the same. The difference between the decision tree and the regression tree is the data type for the nodes. The regression tree uses various libraries. The libraries contain the functions that are needed to perform specific tasks in R. Because we have not covered the data handling processes like data loading, training data selection, testing data selection, feature selection. These all processes helps to convert the raw data into the R compatible file. So, we are directly drawing the trees. The code is explained line by line for better understanding.

library(rpart)

This statement is included at the top of the program because the libraries should be explicitly mentioned at the beginning. rpart library is used for recursive partitioning in survival trees, classification, and regression. It has different functionalities as illustrated below:

- labels.rpart: Creates split labels for a rpart object
- meanvar.rpart: Mean-variance plot for a rpart object
- na.rpart: Handles missing values in a rpart object
- path.rpart: Follows paths to selected nodes of a rpart object
- plot.rpart: Plots a rpart object
- plotcp: Plots a complexity parameter table for a rpart fit
- post.rpart: PostScript presentation plot of a rpart object
- predict.rpart: Predictions from a fitted a rpart object
- print.rpart: Prints a rpart object
- printcp: Displays CP table for fitted rpart object
- prune.rpart: Cost-complexity pruning of a rpart object
- rpart: Recursive partitioning and regression trees
- rpart.control: Control for a rpart Fits
- rpart.exp: Initialization function for exponential fitting
- rpart.object: Recursive partitioning and regression trees object
- rsq.rpart: Plots the Approximate R-Square for the different splits
- snip.rpart: Snips subtrees of a rpart object
- summary.rpart: Summarizes a fitted a rpart object

library(rattle)

Rattle package: This package in R provides a graphical user interface for building and executing the machine learning models. It can be easily installed using the command as follows.
> install.packages("rattle", dependencies=c("Depends", "Suggests"))
The complete working of the Rattle package is described in Section 4.2.

library(party)

It is the toolbox used for recursive partitioning. The base of this package is ctree(). It is used for implementing conditional inference trees. This package can be easily applied to regression problems.

library(partykit)

It provides an infrastructure for summarizing, representing the tree structures for both regression and classification models. All the functions of the party package can also be extracted from this package. This package provides the infrastructure facilities for classification models also, which was not in the 'party' package.

library(caret)

This is the complete toolbox for providing training for the classification and regression models. The functionality of this package includes findCorrelation, filterVarImp, confusionMatrix, varseg, and many more. To describe every functionality is not possible here, so kindly visit the web page https://cran.r-project.org/web/packages/caret/caret.pdf.

data(segmentationdata)

With this command, we are instructing the system to generate some data itself. In this, there are two types of data fetched by the system; PS: poorly segmented and WS: well segmented. The summary of this data is presented in Figure 6.7.

tree.1 <- rpart(form,data=data,control=rpart.control(minsplit=50,cp=0))

The functionality of rpart is used to form a regression tree. You can check the rpart manual for the fitting pfa rpart model. The default command for the same is as follows.
rpart(formula, data, weights, subset, na.action = na.rpart, method, model = FALSE, x = FALSE, y = TRUE, parms, control, cost, ...)

plot(tree.1)
text(tree.1)

```
> summary(data)
 Class        AngleCh1             AreaCh1          AvgIntenCh1        AvgIntenCh2
 PS:1300   Min.   :  0.03088   Min.   : 150.0   Min.   :  15.16   Min.   :  1.0
 WS: 719   1st Qu.: 53.89221   1st Qu.: 193.0   1st Qu.:  35.36   1st Qu.: 45.0
           Median : 90.58877   Median : 253.0   Median :  62.34   Median :173.5
           Mean   : 90.49340   Mean   : 320.3   Mean   : 126.07   Mean   :189.1
           3rd Qu.:126.68201   3rd Qu.: 362.5   3rd Qu.: 143.19   3rd Qu.:279.3
           Max.   :179.93932   Max.   :2186.0   Max.   :1418.63   Max.   :989.5

  AvgIntenCh3       AvgIntenCh4       ConvexHullAreaRatioCh1 ConvexHullPerimRatioCh1
 Min.   :   0.12   Min.   :  0.5633   Min.   :1.006          Min.   :0.5106
 1st Qu.:  33.50   1st Qu.: 40.6797   1st Qu.:1.065          1st Qu.:0.8570
 Median :  67.43   Median : 90.2500   Median :1.149          Median :0.9133
 Mean   :  96.42   Mean   :140.7016   Mean   :1.206          Mean   :0.8958
 3rd Qu.: 127.34   3rd Qu.:191.1704   3rd Qu.:1.281          3rd Qu.:0.9556
 Max.   :1205.51   Max.   :886.8375   Max.   :2.900          Max.   :0.9965
 DiffIntenDensityCh1 DiffIntenDensityCh3 DiffIntenDensityCh4 EntropyIntenCh1
 Min.   : 25.76      Min.   :  1.414     Min.   :  2.307     Min.   :4.708
 1st Qu.: 43.53      1st Qu.: 29.026     1st Qu.: 31.679     1st Qu.:6.029
 Median : 55.81      Median : 54.141     Median : 60.090     Median :6.572
 Mean   : 72.66      Mean   : 74.328     Mean   : 87.183     Mean   :6.578
 3rd Qu.: 79.91      3rd Qu.: 96.858     3rd Qu.:115.064     3rd Qu.:7.038
 Max.   :442.77      Max.   :470.690     Max.   :531.347     Max.   :9.476
 EntropyIntenCh3  EntropyIntenCh4  EqCircDiamCh1   EqEllipseLWRCh1  EqEllipseOblateVolCh1
 Min.   :0.2167   Min.   :0.3734   Min.   :13.86   Min.   : 1.009   Min.   :  146.5
 1st Qu.:4.6120   1st Qu.:4.6667   1st Qu.:15.71   1st Qu.: 1.415   1st Qu.:  277.6
 Median :5.8492   Median :5.8039   Median :17.97   Median : 1.825   Median :  433.3
 Mean   :5.4820   Mean   :5.5067   Mean   :19.48   Mean   : 2.108   Mean   :  714.7
 3rd Qu.:6.6059   3rd Qu.:6.6638   3rd Qu.:21.50   3rd Qu.: 2.450   3rd Qu.:  785.1
 Max.   :8.0128   Max.   :8.2733   Max.   :52.76   Max.   :11.312   Max.   :11689.0
 EqEllipseProlateVolCh1 EqSphereAreaCh1  EqSphereVolCh1  FiberAlign2Ch3  FiberAlign2Ch4
 Min.   :  63.85        Min.   : 603.8   Min.   : 1395   Min.   :1.000   Min.   :1.000
 1st Qu.: 148.36        1st Qu.: 775.7   1st Qu.: 2031   1st Qu.:1.290   1st Qu.:1.260
 Median : 224.36        Median :1014.6   Median : 3039   Median :1.469   Median :1.454
 Mean   : 361.57        Mean   :1283.4   Mean   : 4918   Mean   :1.454   Mean   :1.430
 3rd Qu.: 383.89        3rd Qu.:1452.8   3rd Qu.: 5207   3rd Qu.:1.648   3rd Qu.:1.620
 Max.   :6314.82        Max.   :8746.1   Max.   :76912   Max.   :2.000   Max.   :2.000
 FiberLengthCh1   FiberWidthCh1    IntenCoocASMCh3    IntenCoocASMCh4
 Min.   : 11.87   Min.   : 4.385   Min.   :0.004588   Min.   :0.004514
 1st Qu.: 20.86   1st Qu.: 7.321   1st Qu.:0.009965   1st Qu.:0.017555
 Median : 29.29   Median : 9.608   Median :0.037385   Median :0.049370
 Mean   : 34.72   Mean   :10.274   Mean   :0.099526   Mean   :0.098010
 3rd Qu.: 41.08   3rd Qu.:12.641   3rd Qu.:0.127312   3rd Qu.:0.114280
 Max.   :220.23   Max.   :33.171   Max.   :0.936603   Max.   :0.881854
 IntenCoocContrastCh3 IntenCoocContrastCh4 IntenCoocEntropyCh3 IntenCoocEntropyCh4
 Min.   : 0.01627     Min.   : 0.03444     Min.   :0.2546      Min.   :0.422
 1st Qu.: 4.28495     1st Qu.: 4.05663     1st Qu.:5.0161      1st Qu.:5.101
 Median : 8.48621     Median : 6.42411     Median :6.3618      Median :6.087
 Mean   : 9.92910     Mean   : 7.87706     Mean   :5.8918      Mean   :5.752
 3rd Qu.:13.81686     3rd Qu.: 9.92002     3rd Qu.:7.1103      3rd Qu.:6.805
 Max.   :59.05366     Max.   :61.55833     Max.   :8.0684      Max.   :8.071
 IntenCoocMaxCh3   IntenCoocMaxCh4   KurtIntenCh1      KurtIntenCh3      KurtIntenCh4
 Min.   :0.01429   Min.   :0.01342   Min.   :-1.40260  Min.   :-1.3525   Min.   :-1.5566
 1st Qu.:0.05117   1st Qu.:0.10760   1st Qu.:-0.48851  1st Qu.: 0.1026   1st Qu.:-0.8313
 Median :0.17978   Median :0.21189   Median :-0.00892  Median : 1.4168   Median :-0.3118
 Mean   :0.23196   Mean   :0.24671   Mean   : 0.82479  Mean   : 3.3396   Mean   : 0.9720
 3rd Qu.:0.35331   3rd Qu.:0.33712   3rd Qu.: 0.88109  3rd Qu.: 3.9921   3rd Qu.: 0.7470
 Max.   :0.96833   Max.   :0.94037   Max.   :97.42140  Max.   :99.9808   Max.   :82.7162
 LengthCh1        NeighborAvgDistCh1 NeighborMinDistCh1 NeighborVarDistCh1    PerimCh1
 Min.   : 14.90   Min.   :146.1      Min.   : 10.08     Min.   : 56.89     Min.   : 47.49
 1st Qu.: 22.13   1st Qu.:197.0      1st Qu.: 22.55     1st Qu.: 87.91     1st Qu.: 64.15
 Median : 27.32   Median :227.0      Median : 27.64     Median :107.22     Median : 77.50
 Mean   : 30.36   Mean   :231.1      Mean   : 29.69     Mean   :105.14     Mean   : 89.98
 3rd Qu.: 34.84   3rd Qu.:261.1      3rd Qu.: 34.08     3rd Qu.:121.61     3rd Qu.:100.52
 Max.   :102.96   Max.   :375.8      Max.   :126.99     Max.   :159.30     Max.   :459.77
 ShapeBFRCh1      ShapeLWRCh1     ShapeP2ACh1     SkewIntenCh1      SkewIntenCh3
 Min.   :0.1944   Min.   :1.002   Min.   :1.005   Min.   :-2.6657   Min.   :-1.1136
 1st Qu.:0.5308   1st Qu.:1.329   1st Qu.:1.378   1st Qu.: 0.2823   1st Qu.: 0.8164
 Median :0.6041   Median :1.626   Median :1.777   Median : 0.6093   Median : 1.3049
 Mean   :0.5959   Mean   :1.808   Mean   :2.050   Mean   : 0.6870   Mean   : 1.4892
 3rd Qu.:0.6725   3rd Qu.:2.062   3rd Qu.:2.383   3rd Qu.: 0.9796   3rd Qu.: 1.9166
 Max.   :0.8885   Max.   :7.759   Max.   :8.104   Max.   : 6.8754   Max.   : 9.6690
 SkewIntenCh4      SpotFiberCountCh3 SpotFiberCountCh4 TotalIntenCh1    TotalIntenCh2
 Min.   :-1.0044   Min.   : 0.000    Min.   : 1.000    Min.   :  2382   Min.   :     1
 1st Qu.: 0.4035   1st Qu.: 1.000    1st Qu.: 4.000    1st Qu.:  9500   1st Qu.: 14367
 Median : 0.7283   Median : 2.000    Median : 6.000    Median : 18285   Median : 49220
 Mean   : 0.9325   Mean   : 1.883    Mean   : 6.825    Mean   : 37245   Mean   : 52258
 3rd Qu.: 1.2254   3rd Qu.: 3.000    3rd Qu.: 8.000    3rd Qu.: 35717   3rd Qu.: 72495
 Max.   : 8.0690   Max.   :16.000    Max.   :50.000    Max.   :741411   Max.   :363311
  TotalIntenCh3    TotalIntenCh4     VarIntenCh1      VarIntenCh3        VarIntenCh4
 Min.   :    24   Min.   :    96   Min.   : 11.47   Min.   :  0.8693   Min.   :  2.301
 1st Qu.:  8776   1st Qu.:  9939   1st Qu.: 25.30   1st Qu.: 36.7047   1st Qu.: 47.427
 Median : 18749   Median : 24839   Median : 42.50   Median : 69.1166   Median : 87.251
 Mean   : 26760   Mean   : 40551   Mean   : 72.20   Mean   : 98.5549   Mean   :120.021
 3rd Qu.: 35277   3rd Qu.: 55004   3rd Qu.: 81.77   3rd Qu.:123.8391   3rd Qu.:159.137
 Max.   :313433   Max.   :519602   Max.   :642.02   Max.   :757.0210   Max.   :933.524
   WidthCh1         XCentroid        YCentroid
 Min.   : 6.393   Min.   :  9.0   Min.   :  8.0
 1st Qu.:13.820   1st Qu.:142.0   1st Qu.: 88.0
 Median :16.188   Median :262.0   Median :165.0
 Mean   :17.624   Mean   :260.7   Mean   :177.3
 3rd Qu.:19.784   3rd Qu.:382.0   3rd Qu.:253.0
 Max.   :54.745   Max.   :501.0   Max.   :501.0
```

FIGURE 6.7: Summary of regression data

This command is used to plot the regression tree. The type and number of branches are set by the minsplit argument in the previous command. The text command is used to write the text of values over the different branches.

prp(tree.1)
prp(tree.1,varlen=3)

The pruning is helpful to reduce the complexity of the tree. The least important nodes are split, and the new tree formed contains less number of nodes. Both are trees, i.e., the regression tree (Figure 6.8) and the pruned regression tree (Figure 6.9) are explored with this program.

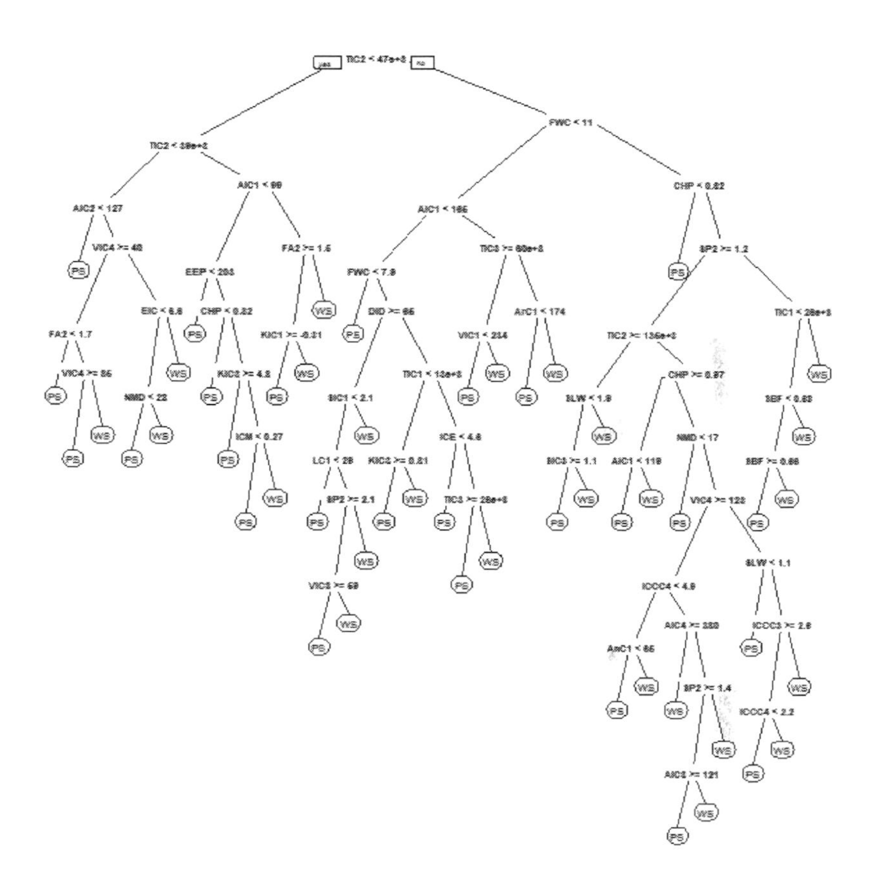

FIGURE 6.8: Regression tree

6.6 Instance-based Learning

There are situations when we want to learn directly by analyzing the instances. The creation of rules is a tedious job, and rules creation becomes more complex when the data is large in volume. This kind of learning is known as instance-

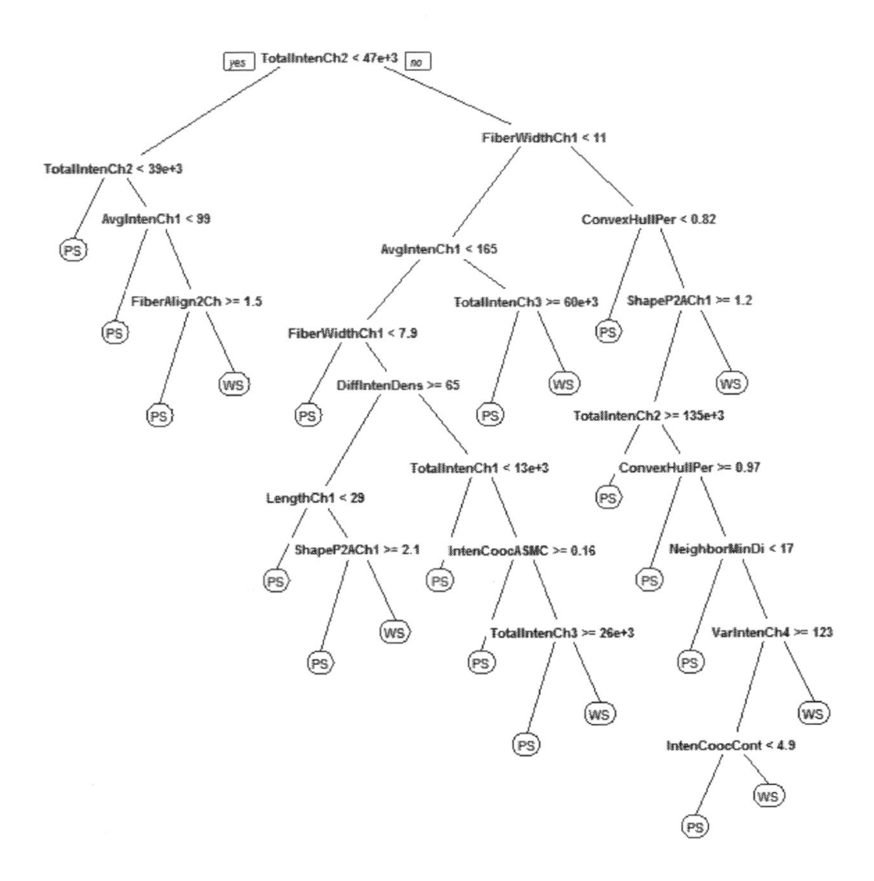

FIGURE 6.9: Pruned regression tree

based learning. The system learns from the instances and represents the results not in the form of a decision tree or rule set but in the sense of instances themselves. It is easy to modify the instances in the data in this learning. The reason behind this is when the learning is rule-based, the new instances should be categorized in one of the rules, but when it is instance-based, the learning is inferred by the instances itself. Whenever the instance is modified or a new one is added then it is compared with the existing instances to search the closest one with the help of distance metric. When a suitable one is found, the new one is placed next to it, forming the nearest-neighbor classification method. What happens when we want to find multiple suitable places for the new instance member? Then the method used is the closest k nearest-neighbors method. The basic idea is the difference, which plays a major role in finding the relationship among the instances. Although it depends upon the number of attributes also. The distance method mainly used is the standard Euclidean distance. This method considers the normalized attributes and their importance as equal. Which attribute is important and which is not; this is still an open research topic. It is biased to compute the distance between the attributes which is done by searching the identical attributes and assigning them zero and one otherwise. Attribute weighting, a distance metric is the basic methods used. It all depends upon the data and its range. The result of this type of learning is that the instances learns itself by discarding the non-trivial and unnecessary instances. The complete framework of instance-based learning is discussed as follows.

1. Learning the different parameters of instances.
2. Finding the relationship among the different instances
3. Selecting low-priority instances.
4. Discarding the selected the instances.
5. Forming the new data with the new structure.

It can't be said that it is the correct kind of learning. So, it does not lie in any of the categories of supervised and unsupervised learning. This learning does not target finding the hidden patterns in data, which is the primary objective of machine learning. It targets finding the outliers by learning with the mechanism of distance.

6.6.1 Distance Metric

In instance-based learning, the new instance searches for the nearest member of any class. Once the class of the nearest member is known, the same class is assigned to the new instance. The distance between the two instances can be calculated using Euclidean distance formula, in which n is the number of instances.

$$X = \sqrt{(a_1^1 - a_2^1)^2 + (a_1^2 - a_2^2)^2 + (a_1^n - a_2^n)^2} \tag{6.1}$$

Equation 6.1 may lead to undesirable differences when the difference is very large. So, it is very necessary to normalize the values between a particular range, which can be calculated with the formula given below:

$$a^i = \frac{n^i - min(n^i)}{max(n^i) - min(n^i)} \tag{6.2}$$

In Equation 6.2, n^i refers to the actual value of the instance. Min and max are the minimum and maximum value, taken as 0 and 1, respectively.

Instance-based learning is implemented using the dataset of IoT. Figure 6.10 presents the distance computed among the various instances. Figure 6.11 shows the classes formed with the four features, namely wet bulb temperature, total rain, wind speed, maximum wind speed.

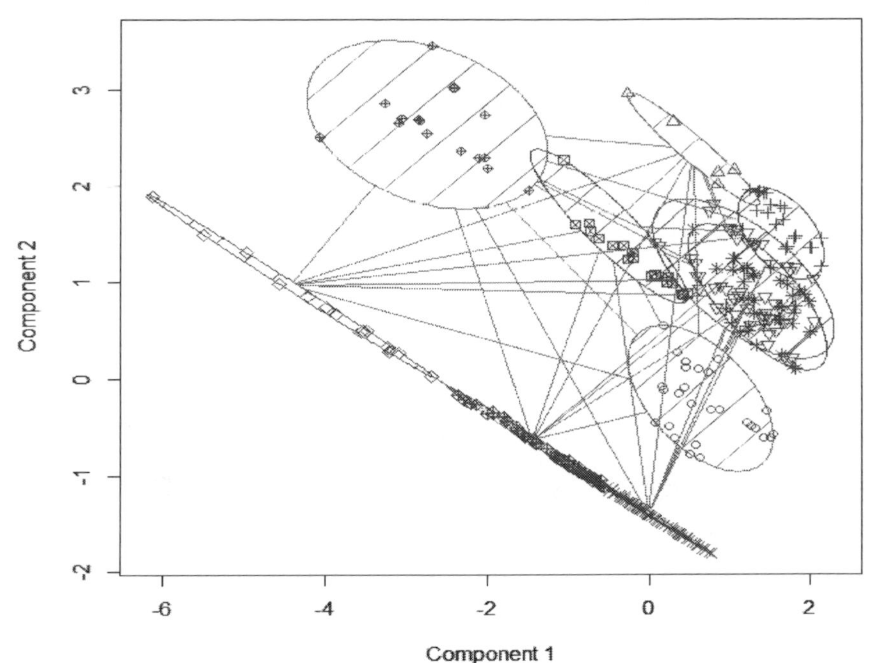

FIGURE 6.10: Distance among the various instances

6.7 Summary and What's Next

This chapter describes the methodologies used in machine learning. It explains that how an instance becomes the input for the machine learning model and how the output becomes an instance. This chapter also illustrates that how the machine learning model takes the decision in case of multiple situations. These issues become an integral part of this chapter. The classification rules used by the machine learning models were discussed. The machine learning model training with the number, instance, coordinates, was also explained. Understanding these concepts helps the learner develop a machine learning model.

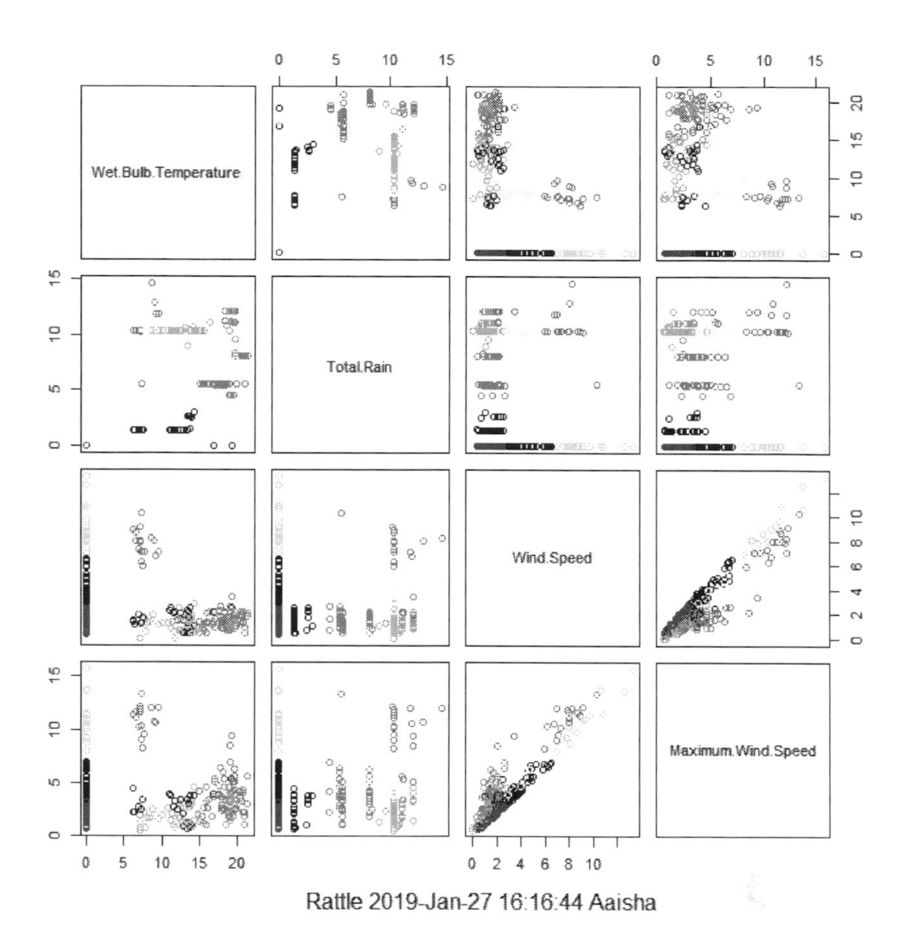

FIGURE 6.11: Classes formed after matching the desired instances

6.8 Exercises

Questions

1. What are the different input parameters to feed the data in a system?
2. Describe the working of a decision tree in detail.
3. What is meant by classification rules?
4. What is replication? How can you overcome this problem?
5. How can you predict numerical values using machine learning?

6. What is meant by instance-based learning?
7. Define distance metric.
8. Describe a replicated subtree in detail.
9. Explain the rpart library with in-built functions.
10. How can you represent data in the form of structures?

Multiple Choice Questions

1: Which kind of data is used for the formation of a flat file?

(a) Rows and columns
(b) Image dimensions
(c) Raw data
(d) None of these

2: The data in the data structure tree is categorized by:

(a) Dimensions
(b) Relationships
(c) Conditions
(d) None of these

3: Classification is a processs of splitting data on the basis of:

(a) Classification decisions
(b) Classification conditions
(c) Classification rules
(d) None of the above

4: How the output are generalised after being predicted by machine learning model?

(a) Decision table
(b) Decision tree
(c) Decision rules
(d) None of the above

5: What is the role of the rpart library?

(a) Recursive partitioning
(b) Classification
(c) Regression
(d) All of these

6: What is the use of rattle library?

(a) Graphics
(b) GUI
(c) Decision table
(d) None of the above

7: Is the party library used for recursive partitioning in conditional inference trees?

(a) True
(b) False
(c) All the decision trees
(d) None of the above

8: What is the difference between party and partykit library?

(a) Infrastructure capability
(b) Representing capability
(c) Classification
(d) Regression

9: Which library can implement all the machine learning models?

(a) party
(b) rpart
(c) caret
(d) None of above

10: Which capability of R allows to split the data by the form of segments?

(a) segmentationdata library
(b) segment function
(c) split function
(d) None of the above

11: Which function is used to draw the regression tree?

(a) reg()
(b) rpart()
(c) rnode()
(d) None of the above

12: What is the role of instance-based learning?

(a) Finding hidden patterns
(b) Detecting outliers
(c) Data reduction
(d) None of these

13: Which method is used to find the distance between the instances while computing the distance metric?

(a) Nearest neighbor
(b) Euclidean distance
(c) Mean of difference
(d) None of the above

14: The problem of replicated subtree is due to the:

(a) Un-analyzed conditions
(b) Large dataset
(c) Presence of anomalies
(d) All of the above

15: What should be done if the instance does not belong any category of rules formed?

(a) Revise the rules
(b) Change the parent node
(c) Add a new rule
(d) None of the above

16: Which function of rpart library handles missing values?

(a) na.rpart()
(b) predict.rpart()
(c) path.rpart()
(d) None of the above

17: Which function of rpart library helps in prediction?

(a) rpart.predict() (b) predict.rpart()

(c) na.rpart() (d) path.rpart()

18: What is the full form of rpart?

(a) Recurrent partitioning and regression trees (b) Recursive partitioning and regression trees

(c) Recursive partitioning and replicated trees (d) None of the above

19: Why the instance based learning is adopted?

(a) Relationships between instances (b) Outlier detection

(c) Low priority instances (d) All of the above

20: In case of instance based learning, why do we discard few instances?

(a) Outliers (b) Low priority instances

(c) Normalization (d) All of the above

7

Different Machine Learning Models

The learning discussed in the previous chapter is tree-based. It works well with nominal attributes. In a tree structure, the numeric prediction is achieved by applying a few tests into the classification rules. The direct implementation of numeric values in the tree is restricted. However, there are other learning methods which work directly with the numerical attributes. Now, in this chapter, we learn about those methods (models), their methodologies for working with respect to various components.

7.1 Linear Method for Regression

When all the attributes including the target attributes are numeric, then the linear regression comes into the action. It is the old method in statistics. The idea is to represent a class with the attributes after being supplied with the weights. The equation used is as follows. In Equation 7.1, a is the class, x_k are the instances, and w_k are the weights computed by a model.

$$a = w_0 x_0 + w_1 x_1 + w_2 x_2 + ... + w_n x_n \tag{7.1}$$

In the training data, every instance has the value for every attribute, which forms a class. In the case of linear regression, every class also carries the weight of each instance. So, the equation can be written as follows.

$$a^1 = w_0 x_0^0 + w_1 x_1^1 + w_2 x_2^2 + ... + w_n x_n^n \tag{7.2}$$

In this Equation 7.2, the a^1 is the actual class, and the right hand of the equation, which is the summation of weights with its instance, is the predicted class. Now, the main consideration is for the target which is expected to be outcome of machine learning model. The difference between the actual and the predicted should be minimal.

$$\sum_{j=0}^{m} \left\{ a^j - \sum_{i=0}^{n} w_i x_i \right\}^2 \tag{7.3}$$

In Equation 7.3, the difference between the sum of squares of actual classes and predicted classes are computed. There are m training instances with n attributes.

7.2 Linear Method for Classification

The linear method for classification can be formed by the linear method for the regression. Only the constraint is to fix the output to one in case the instance belongs to the class, otherwise it is zero as shown in Figure 7.1. There is *multi-response linear regression* also, in which the instance belonging to the class leads to multiple outputs. But like the other models, they also have a

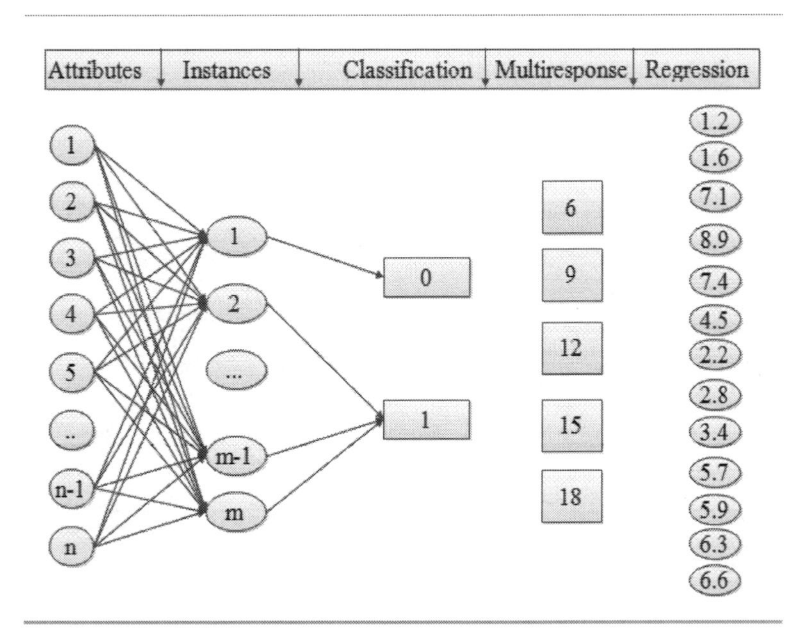

FIGURE 7.1: Linear model of classification, multi-response, and regression

disadvantage that is it can only represent linear boundaries between different classes. To overcome this problem **support vectors machines** (mainly known as support algorithms) implements linear models to enforce non-linear class limits where a linear border may not separate classes.

How does it work? It transforms the input using a non-linear mapping which means it transforms the space into a new space. On completion, the new model is constructed in a new space and represents a non-linear decision boundary in the original space. It can be done by letting the original set of attributes replaced by 1 by giving all products of n factor. Consider the following example to understand more clearly.

1. Take two attributes, x_1, x_2.

2. Let a be the outcome.

3. Let w be the weights.

4. Consider three factors of all the products.

The equation formed is as follows:

$$a = w_1 x_1^3 + w_2 x_1^2 x_2^1 + w_3 x_1^2 x_1^2 + w_4 x_2^3 \tag{7.4}$$

Now, each training instance is mapped into a new space to generate the model in the new space by computing all possible three-factor products of its two attribute values. The outcome can be used by training one linear system for each class for classification. After which we can assign an unknown instance to the class that gives you the greatest output in a. There are a couple of drawbacks to this approach.

1. A large number of coefficients are generated by the transformation in any realistic setting. For example, consider 10 attributes as the original dataset. Let us assume that they include 5 factors each. Then the learning algorithm determines around 2000 coefficients. It leads to the complexity of the algorithm.

2. The resulting model is also non-linear if the number of coefficients is large due to the large number of training instances. So, it results in overfitting the data. (By overfitting, we mean the production of an analysis that corresponds closely or exactly to a particular set of data which may fail to fit additional data or predict future observations reliably.)

What are the possible solutions we can think of?

Here, support vectors aid in solving these problems. They find a different kind of linear model, which is called the maximum margin hyper-plane. To understand what a maximum margin hyper-plane is, imagine a two-class dataset whose classes are linearly separable. If this is the case, then there exists a hyper-plane which is efficient to classify all training data correctly. It gives the greatest separation between the classes.

In geometry, hyper-plane separation theorem is a theorem about disjoint convex sets in n-dimensional Euclidean space. There are several rather similar versions. In one version of the theorem, if both these data sets are closed (a set is closed if and only if it coincides with its closure) and at least one of them is compact (compactness is a property that generalizes the notion of a subset containing all its limit points and bounded), then we can say that there is a hyper-plane in between the sets and even two parallel hyper-planes in between them separated by a gap. Consider the Figure 7.2, in which a represents the maximum margin hyper-plane and b represents the support vectors. Observations:

1. The outline appears when you connect points of the set.

2. Convex null cannot overlap in linearly separable classes.

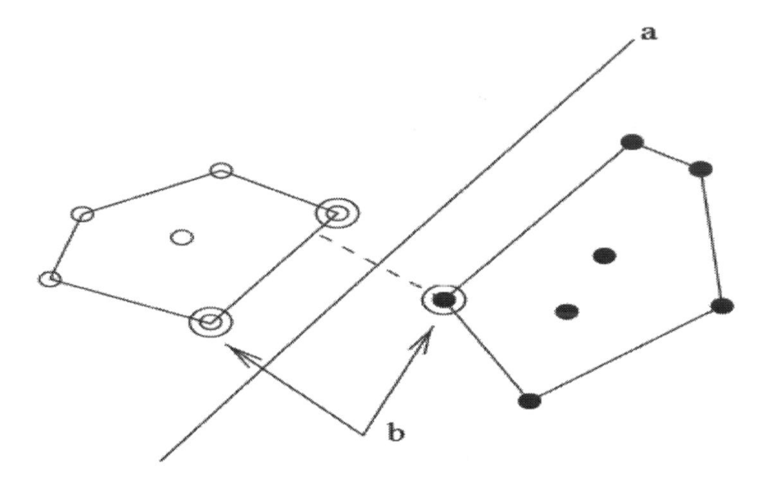

FIGURE 7.2: Maximum margin hyper-plane

3. The highest hyper-plane margin is always the one that is as far away from both convex hulls as feasible.

4. There is at least one support vector in each class.

5. The set of support vectors here describes the maximum hyper-plane margin for the learning issue in a unique way.

It is important to note that if we are supplied with support vectors for the two classes then we can build the maximum margin hyper-plane, and the other training data instances can be removed without changing the hyper-plane's orientation. The new equation can be formed after removing the dataset elements.

$$a = w_0 + w_1x_1 + w_2x_2 \tag{7.5}$$

Here, if we take support vector in consideration, the equation can be modified as follows.

$$a = b + \sum p_i y_i x(i).x \tag{7.6}$$

where,

- y = Instance class value y as either 1(if yes) or -1(if no). This is the x(i) example of practice.
- x= test instance
- x(i)= support vectors
- x(i).x= dot production notation
- $p_i y_i$ = parameters that define the hyper-plane.

Problems? To find $p_i y_i$ leads to a problem known as constrained quadratic optimization which can be solved using software packages. Lastly, computational complexity can be decreased and learning can be accelerated by applying distinctive algorithms to support vectors for practice. On considering the case of linear class boundaries it leads to computational complexity and overfitting. If we use support vectors for help, the chances of overfitting are comparatively reduced. This is performed by altering one or two instance vectors that create sweeping adjustments to the decision boundary's big parts. On the other side, the hyper-plane is very stable. It only passes when training cases that are support vectors as stated above are added or deleted. (This is true in the case of non-linear transformation of high-dimensional space). Taking into consideration, computational complexity is optimized by calculating the dot product before the non-linear mapping is done on the original attribute. The optimized equation can be written as follows.

$$a = b + \sum p_i y_i (x(i).x)^n \tag{7.7}$$

Where n is the conversion factor number. A useful way to select the value for n is to begin with 1 (a linear model) and increase it until there is no improvement in the projected error. It's enough to have a fairly tiny value. A polynomial kernel function $(x.y)^n$, which calculates the dot product of two vectors x and y and increases the result to power n, can be used for different implementation of non-linear mapping. The radial-based kernel function and the sigmoid kernel are two additional features used for execution. The support vector with radial base function kernel is a sort of neural network called an RBF (radial base function) network, and the support vector with sigmoid kernel implements another type of neural network, a multilayer perceptron with one hidden layer. Even the fastest training algorithms to help vector machines become slow for a non-linear environment.

7.3 Kernel Smoothing Models

We can use the perceptron kernel algorithm to learn the boundaries of non-linear decisions. In this case, instead of the dot product, we may use a kernel function. Therefore, a non-linear classifier can be learned by the perceptron algorithm simply by keeping track of instances that were misclassified during the training process and using this expression to form each prediction. This can also be implemented to transform it into a logistic regression kernel. The solution acquired is not scarce, i.e., the solution vector is supported by every training example. Using a weighted vector is one way to make it more stable. After its beginnings, the number of consecutive paths can lead to the, 'correctness,' of a weight vector in which it properly categorized subsequent instances. The voted perceptron is the name given to this algorithm version that can be used as the number of votes given to the weight vector, which performs nearly as well as a support vector machine.

7.4 Back Propagation

In the case of a multilayer perceptron, which is a predictor of the underlying classification problem, we need to determine the appropriate weights for the network connections. When there are no hidden layers, we can use a rule of learning to do this. If there are two hidden layers, it is possible to apply a similar strategy a second time to update the current weights of the first hidden layer's input connections, propagating the error from the output unit to the first through the second hidden layer. The generic gradient descent approach is known as backpropagation because of this error propagation mechanism. But the algorithm fails to predict the correct output in the case of hidden units, and because they are hidden, there the rule cannot be followed.

Solution

This issue is solved by modifying the weights that lead to the hidden units. Based on the intensity of the contribution of each unit to the final forecast, they can be altered. Gradient descent is a mathematical method which helps to achieve this. Taking derivatives is the strategy that follows that includes the step function used to transform the weighted sum of the inputs into a 0/1 forecast that the perceptron does not distinguish.

The error function to be minimized by changing the weights should be differentiable to apply the gradient descent procedure. It's not always useful to use this procedure. It can take advantage of the data provided by the function derivative to be minimized.

Example: Consider a hypothetical error function that is identical to x2 + 1. The x-axis that represents a hypothetical parameter is to be optimized. x2 + 1 derivative is just 2x.

The resulting observation is based on the derivative. We can also figure out the function's path at any given stage. If the slope of the feature is down to the right, its derivative will be negative; if it is down to the left, it will be positive. The size of the derivative determines the steepness of the decrease. This data is used by gradient descent to adjust the parameters of a function that is an iterative optimization method. It requires the derivative value, multiplies it by a tiny constant called the learning rate, and subtracts the outcome from the present value of the parameter. As it is iterative, it is repeated for the new value of the parameter, and so on until it reaches a minimum. The learning rate determines the size of the step and the speed of convergence of the search.

If it has multiple minima and is too big, a minimum value may be missed by the search, or wildly processed. If it is too low, progress towards the min-

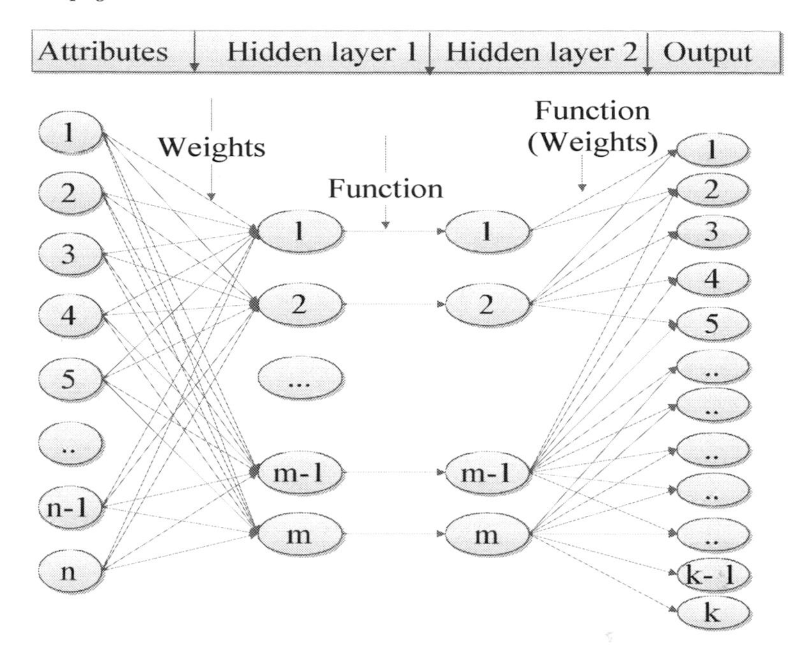

FIGURE 7.3: Multilayer perceptrons

imum may be slow. Only a local minimum can be found for the gradient descent. If multilayer perceptrons generally have multiple minima features for the function and error features, it may not discover the best one. This is an important disadvantage of conventional multilayer perceptrons compared to supporting vector machines, for instance. It is possible to obtain a more compact classifier from a single network to create an output unit for each class that links each unit in the concealed layer to each output unit as shown in Figure 7.3. Multilayer perceptrons belong to a network class called the feedforward network because they contain no cycle and the output of the network depends only on the current input instance. Once all training instances are supplied through the network and all weights are gathered, only the update takes place. As all training data are processed together, this phenomenon is known as batch learning.

Stochastic backpropagation is incrementally weight updates after processing each training example. It is used for online learning where a continuous stream of new data arrives and every training instance is processed only once. In the methods of propagation, zero mean and unit standard deviation as always helpful.

Overfitting is the disadvantage of multilayer perceptrons. The drawback of multilayer perceptrons are that they include fundamentally opaque hidden units.

The rule based learners operates as a reduced-error pruning method to prevent overfitting. In other words, it functions as a holdout set that is used to decide when to stop using the backpropagation algorithm to perform further iterations. Another technique that can be used is called weight decay, which adds a penalty word to the error function consisting of the squared amount of all network weights.

7.4.1 Radial Basis Function Networks

This is a feed-forward network that contains two layers of computation performed by hidden units. For the specified example, the hidden device functions as activation. If the instance is closed or vice versa, the activation can be strong. Here, hyper-sphere replaces the hyperplane. Hypersphere is the space for which the same activation is produced by a given hidden unit. The output layer takes a linear combination of the hidden unit's outputs and pipes it through the sigmoid function in classification problems. An important benefit over multilayer perceptrons is that it is possible to determine the first set of parameters independently of the second set and still generate precise classifiers. There are distinct techniques for determining the first set of parameters. One technique is to use clustering without looking at the training instances, class labels at all. It is possible to apply the easy k-means of the clustering algorithm, which clusters each class separately to acquire k basic features for each class. On the other side, it is possible to learn the second set of parameters while maintaining the first parameters. RBF networks have a disadvantage so they offer the same weights to each attribute as all are treated similarly when the distance is calculated.

7.5 Neural Network

The neural network is recognized as a network modeled after the activity of neurons in the human brain. It is a scheme of computing that follows a procedure. It begins with the first line of code and proceeds to the next line. It follows linear fashion directions. A real neural network never follows a linear route, but the data is jointly processed in parallel throughout a node network. The main component of the neural network is its learning capacity. It is a complex adaptive system that, depending on the data flowing through it, can alter its inner composition. It is accomplished by weight adjustment. The network components are known as neurons, which are easy to understand. They read an input, further process it, and produce the appropriate output.

Each row represents a link between two neurons and the pathway for information flow is further indicated. Each link has a weight and a number that enables the two neurons to regulate the signal. If the resulting product is not appropriate the weights are adapted.

Standard uses of neural networks
1. Pattern recognition
2. Time series prediction
3. Signal processing
4. Soft sensors
5. Control
6. Anomaly detection

7.5.1 The Perceptron

The perceptron is regarded as the easiest neural network. This is a single neuron's computational model. It is comprised of a single output, one or more inputs, a processor. It follows a model of a feed-forward network, which means the inputs are sent to the neuron and processed and the outcome is produced. In the list below, we can see that the network reads the input coming in and going out from left to right.

Procedure:

1. Receive inputs
   input 1=12
   input 2=4

2. Weight inputs
   Each input in the neuron should first be weighted, i.e., multiplied by some value (the number is often between -1 and 1)
   Here, weight for input 1=0.5 and weight for input 2=-1
   So the calculated input will be
   input 1=12*0.5=6
   input 2=4*-1

3. Sum inputs
   Sum=6

4. Generate outputs
   The output is produced by passing the amount through the activation function. If the sum is the positive number, the output is one and vice versa, we can make the activation function the sign of the sum, i.e., if the sum is the positive number, the output is one and vice versa.

7.6 Bayesian Methods

Bayesian methods use the data analysis process to reduce the property of the underlying probability distribution. Bayes theorem is used to update the probability of a hypothesis as information becomes more available. Sequential analysis is used to update the data. The equation is stated as below:

$$P(x|y) = \frac{P(y|x)P(x)}{P(y)} \tag{7.8}$$

which is known as Bayes' rule. Frequent statistics test the occurrence of an incident or not. It calculates the probability of an event in the long run of the experiment. The sampling distribution of the specified size will be done under this. Then the experiment is repeated theoretically an infinite number of times. For instance, when the experiment is repeated 1000 times, there must be some stopping criteria. For example, repeat until we get 300 heads in a coin toss. It has a very prevalent error, i.e., the reliance on the number of times the experiment is repeated from the consequence of an experiment.

1. P-values are measured against a set size of statistics and with an intention to change sample size. For example, when the total number reaches 100, Person A may choose to stop tossing a coin while B stops at 1000. Here we get distinct t-scores and distinct p-values for distinct sample sizes. Likewise, the intention of stopping may change from a fixed number of flips to a total flipping duration. We're also bound to get different p values in this case.

2. Sample size relies strongly on the confidence interval (CI) like p-value. This makes the stopping potential ridiculous because the findings should be compatible regardless of how many individuals perform the trials on the same information.

3. Trust intervals (CI) are not distributions of probability, so they do not provide the most likely value for a parameter.

7.6.1 Bayesian Statistics

Bayesian statistics are regarded as a mathematical operation that relates probabilities to statistical issues. Suppose Rian won three times out of all four championship races (F1) between Rian and John, while John only managed 1. So, if you'd bet on the next race winner, who'd bet on him? You would say, Rian. This is the twist. What if you're told it rained when John won and once when Rian won, and the next day it's definitely going to rain. So who are you going to bet on now? It's simple to see that there has been a huge increase in chances of winning for John. But how much is the issue? To understand the problem, we shall become a little familiar with some concepts, first of which is a conditional probability.

7.6.2 Bayesian Inference

To comprehend the concept behind Bayesian inference, take an instance of coin tossing. A significant aspect of Bayesian inference is the establishment of parameters and models. (Models are the mathematical formulation of the observed occurrences and variables in the models influencing the observed information are the parameters). Now, in tossing a coin, fairness of coin is defined as the parameter of coin denoted by θ. And let the outcome be denoted by D. What is the probability of 4 heads out of 9 tosses(D) on given the fairness of coin (θ), i.e., $P(D|\theta)$.

Given an outcome (D), what is the probability of the coin being reliable (θ=0.5) We represent it using Bayes theorem as follows.

$\mathrm{P}(\theta|\mathrm{D})=(\mathrm{P}(\mathrm{D}|\theta) \text{ X } \mathrm{P}(\theta))/\mathrm{P}(\mathrm{D})$.

Here, P(θ) is the prior (in other words the strength of our belief in the fairness of coin which is before the toss). It is okay to believe that coin can have any degree of fairness between 0 and 1.

P(D|θ) is the observing result given the distribution for. If we knew that coin was fair, it will give the probability of observing the number of heads in a defined number of flips. P(D) stands for the evidence. It is the probability of a data as determined by summing (or integrating) across all possible values of θ, weighted by how strongly it is in those particular values of θ. P(θ|D) is the posterior belief of the parameters after observing the number of heads. To define the model correctly, we need the two mathematical models before it. The first represents the likelihood function P(D|θ), and the second is the distribution of prior beliefs. Also, the product of these two gives the posterior belief P(θ|D) distribution. Since prior and posterior, both are the beliefs about the distribution with the fairness of coin, assuming that they will have the same mathematical form.

There are different algorithms, but the simplest one is k2, which begins with the attributes or nodes being ordered. In turn, it processes each node and considers the edges to the present one from earlier processed nodes. In each step, it brings the edge of the previously processed nodes to the current one. It adds the edge, which maximizes the rating of the network. If there is no further enhancement scope, then consideration is given to the next node. It is possible to limit the number of parents for each node to a fixed value defined by the maximum value. Since only the edges are processed in advance, and it has a fixed order, cycles cannot be introduced. Although the outcome depends on the original order, running the algorithm several times with distinct random orderings make sense.

The naïve Bayes classifier is a network with an edge that leads to each of the other characteristics or nodes from the class attribute. It helps to use this network as a starting point for searching when constructing classification networks.

We can do this in the K2 algorithm by using the class variable to be the first in ordering and appropriately initializing the set of edges. Another trick is to make sure that in the Markov blanket of the node that represents the class attribute, every attribute should be covered.

A node involves all parents, kids, and parents of kids. A node is observed to be conditionally independent of all other nodes provided. If a node is missing from the Markov blanket of the class attribute, the value is meaningless to the classification.

If K2 discovers a scheme that excludes a significant characteristic in the Markov cover of the class hub, it may include an edge that rectifies this deficiency. A simple way to do this is to include an advantage from the centre of the credit to the centre of the class or from the centre of the class to the centre of the property, depending on which decision is kept away from a cycle. However, a gradually contemporary yet slower type of K2 is not to arrange the hubs to voraciously consider including or erasing edges between discretionary hubs sets (while, of course, ensuring acyclicity). We can also consider reversing the heading of current edges. Similarly, as with any insatiable calculation, the following scheme only speaks to a near limit of scoring ability: it is constantly appropriate to run such calculations several times with different uneven introductory settings. Progressively sophisticated methodologies of enhancement can also be used, for instance, mimicking reinforcement, tabu pursuit, or hereditary calculations. For the Bayesian system classifiers, another learning calculation is called naïve Bayes (TAN) extended tree.

It takes the naïve Bayes classification, as the name suggests, and adds edges to it. The class characteristic is the single parent of each naïve Bayes hub arrangement: TAN is considering adding a second parent to each hub. If the class center and any comparing edge are prevented and there is effectively one hub to which a second parent is not included, then the later classifier will have a tree structure formed at the parentless hub – this is the location from which the name comes.

There is an efficient calculation for this limited type of system to find the arrangement of edges that boosts the likelihood of the system dependent on processing the most extreme weighted spread over a tree of the system. All the scoring measurements we have shown so far are likelihood located as they are meant to increase the joint likelihood of Pr[a1, a2,...,aN] for each example.

7.7 Summary and What's Next

In this chapter, the working of various machine learning models was explained in details. The learning of problems such as regression and classification were discussed in the previous chapter. This chapter focuses upon the solutions for these problems. The algorithms such as kernel perceptron algorithm, the neural network are also discussed. A detailed description of the Bayesian method along with its statistics and inference were discussed.

7.8 Exercises

Questions

1. Detail the linear method for regression.
2. What is the difference in the nature of the linear method for classification and regression?
3. Explain the working of SVM.
4. What are the kernel smoothing models?
5. Explain neural networks in detail.
6. Elaborate on the concept of perceptrons with an example.
7. Introduce the Bayesian methods.
8. What is meant by Bayesian statistics and Bayesian inference?
9. Discuss the K2 method in detail.

Multiple Choice Questions

1: The linear method for classification is adopted when the target output is:

(a) Zero (b) One

(c) Both (d) None of these

2: The output of the linear method represents:

(a) Linear values (b) Linear boundaries

(c) Linear space (d) All of them

3: Mapping performed by SVM is:

(a) Linear **(b)** Non-linear

(c) Cannot be defined **(d)** None of the above

4: Mapping performed by the kernel model is:

(a) Linear **(b)** Non-linear

(c) Cannot be defined **(d)** None of the above

5: The different layers in the multilayer perceptron are connected with:

(a) Weights **(b)** Signals

(c) Activation function **(d)** None of these

6: The system convergence can be estimated with the help of:

(a) Training ratio **(b)** Learning rate

(c) MAE **(d)** None of the above

7: Radial basis function returns the points in the form of:

(a) Hyper-plane **(b)** Hyper-sphere

(c) Action function **(d)** None of the above

8: For what purpose are feed forward neural networks primarily used?

(a) Classification
(b) Feature mapping
(c) Pattern mapping
(d) None of them

9: Time series prediction uses:

(a) SVM
(b) Neural network
(c) Kernel smoothing model
(d) None of above

10: Which model is followed by the perceptron?

(a) Back propagation model
(b) Feed-forward model
(c) Kernel smoothing model
(d) SVM

11: The Bayesian method follows:

(a) Bayes, theorem
(b) Bayes, rule
(c) Bayes, function
(d) All of the above

12: The statistics followed by the Bayesian networks apply:

(a) Probabilities
(b) Algebra
(c) Bayes, theorem
(d) All of the above

13: How can the compactness of the Bayesian network be described?

(a) Fully structured

(b) Partial structure

(c) Locally structured

(d) All of the above

14: What is the consequence between a node and its predecessors while creating the Bayesian network?

(a) Dependant

(b) Conditionally independent

(c) Functionally dependent

(d) a & b both

15: What do we mean by the learning rate of networks?

(a) Step size

(b) Weights of network

(c) Convergence power

(d) All of the above

8

Data Processing

Before moving to the working of a machine learning model, we need the data. The data should be clean and noise-free. This data works as an input for the machine learning model. Preparing the input for further investigation requires a lot of efforts. It is worth noting the various complexities that arise while preparing the input for the machine. As we know that the machine tends to learn in the machine learning process, providing the best input leads to better learning.

8.1 Input Preparation

It is necessary to bring the data at one place, to form a particular number of instances. This phase is more commonly known as data collection. But we are not only in the process of collecting the data but need the data to be clean and noise-free. The gathering of data may vary from people to people according to their profession. There are various types of the audience by whom the data is produced and required. If I talk about students, the student is the one who looks for very simple task and attempts to download the data from the Internet. There are many advantages and disadvantages to this analyses. Firstly, no doubt there are hundreds of repositories on the World Wide Web (WWW), like uci, data.science, and many more. Secondly, it is good to download the data because the data is prepared for such experiments. The issues are that the data is not real and up to date. It may have been collected years ago. The experiments being performed by the students would neglect them from learning an important phase of data preprocessing. From a real-world business point of view, data is gathered from various departments of an organization. There can be various departments such as production unit, billing department, sales department, advertisement department, and many more. Most importantly, the data is produced by IoT application. IoT data is gathered with the help of IoT devices. The data collected by an IoT file is large in volume and of different formats. The IoT data have different formats such as BDB format and graphic format. When we collect the data from various sources, there are various challenges. It may lead to different issues such as the different style of recording, different time periods, different data formats, different aggregations, and most importantly different types of errors. Although the data is collected with efforts but still processing is required once it is assembled and gathered. This is what actually leads to data warehousing. The data warehouse is the place where corporate data is stored in years, used to make business decisions. The consistency maintained is the same as the database management systems. This is the immense example of observing the fragmented data of day-to-day operations of industry, being gathered to predict the information equivalent to the gold coins for the industry. The data aggregation must be done by knowing its purpose. For example, weather scientists gather the different parameters of data from various sources. But the data aggregation and denormalization are done in such a manner that the essential parameters are focused on that which are actually required for forecasting. Below are the different issues to be taken care of while data gathering and data aggregation of the input for a machine learning model.

1. File format: There are various file formats we can execute where data is collected and stored. The data file may have attributes and instances. The reason of different file formats is due to the different relationship among the instances and its attributes.

    - ARFF: It is the standard way of representing the datasets. The data in this file is independent, and the instances are not ordered. This leads to no relation among the instances. Referring to the example in Section 5.2, the ARFF file of weather data is presented in Figure 8.1

```
ARFF file of weather data
%
$ attribute outlook {Sunny,Cloudy,Rainy,Moderate}
$ attribute temperature numerical value
$ attribute Atmosphere {Hot,Cold,Normal}
$ attribute Wind {Yes,No}
$ attribute Humidity numerical value
$ attribute Day {Any week day from sunday to saturday}
7 instances
%
Sunny 87 Hot No 90 Sunday
Cloudy 67 Cold Yes 80 Monday
Rainy 74 Cold No 65 Tuesday
Moderate 80 Normal Yes 75 Wednesday
Moderate 82 Normal No 90 Thursday
Sunny 94 Hot No 90 Friday
Cloudy 72 Cold No 90 Saturday
```

FIGURE 8.1: Layout of ARFF file

TABLE 8.1: Supermarket sale record

Customer Id	**Drinks**	**Sweets**	**Spices**	**Stationery**	**Cleaning**	**Cost**
AB4	0	0	0	Pen	0	10
GH6	Frooti	0	0	0	0	50
GR6	0	0	0	Notebook	Surf	450
YH5	0	Chocolate	Dal	0	0	10
12D	Coke	0	0	0	0	100
TR1	0	Sugar	0	0	0	78

presents the weather data. You can see the comments with %. But the data does not have an attribute defined. You have read the comments before preceding for processing this file. Every block of comment in which an attribute is defined has defined the possible values for instances. This format is rough as well as not in the appropriate layout. This file does not define the purpose of the data and does not define the size or dimensions of the dataset. The numeric value or the string value is declared within the comment. Every instance is not headed by the attribute.

When the data is collected from various sources, then many files are combined together to form a zip folder, mainly with .jar extension.

- Sparse data: The data is in the format when it contains a maximum of zeros, rather than the decimal value. Dealing with the zero values is again a very tedious job. For example, data of a supermarket in such a manner that it stores the items purchased by the customer in an instance. The items purchased are marked in each row. For example, look at Table 8.1.

  You can see there are maximum values as zero. It leads to the handling of zero values. Sometimes, the system may handle the zero values by '?', which is again trouble.

- BDB formats: The IoT data mainly takes the form of the binary data buffer (BDB). This format has the most important characteristics of transferability and being self-describable. It takes formats like the XML data model (XBDD), JSON data model (JSONB), CSV data model (CSVB), and HTML data model (HBDB). The system should have java installed to read such formats.
- Graphic formats: The maximum number of IoT devices capture images. These images are analyzed for the extraction of features. The advanced technology deep learning can be used to read the data from images. The bitwise image processing can also be used for the same.

2. Handling the missing values:
   There can be a situation when data is missing from the data after collecting it. As observed in the IoT dataset used in the previous chapters also, the data was completely missing from two rows. The layout of data on the website is presented in Figure 8.2. The method we adopted for handling this number of data rows is deleted. We simply deleted these rows because the rows did not contain any data. So random values cannot be inserted which are not aware of range, type, and a class of data. There are various ways of handling missing values. A few of these are discussed below.

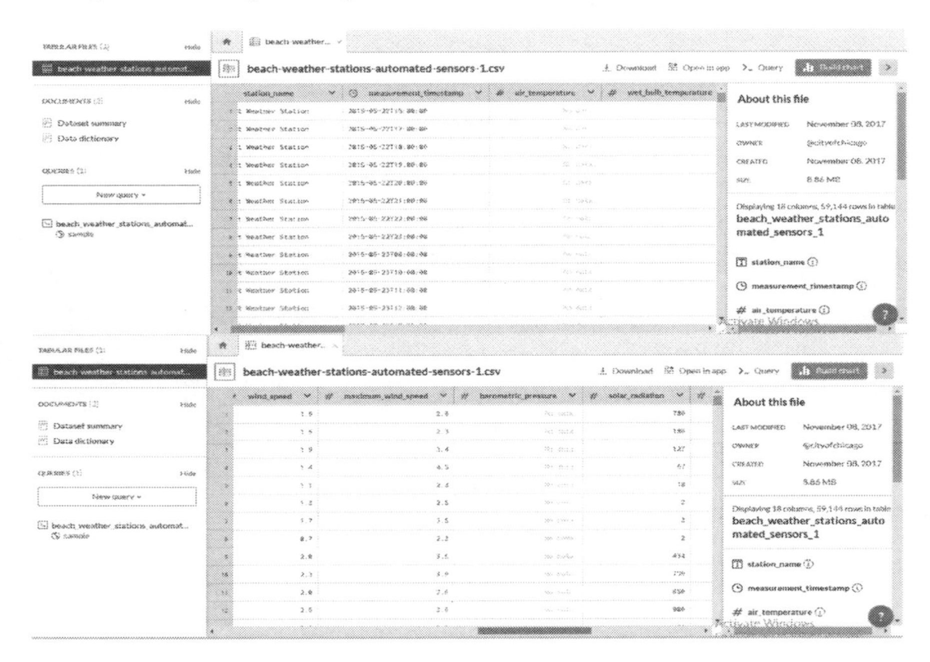

FIGURE 8.2: Rows without data in the IoT experimental dataset

- Deletion: The very first solution for missing values is to delete the row if it contains more than 30% of missing values or delete the column if it contains 50% of missing values. This is the right solution if such a situation occurs. But there are various things to be kept in mind while adopting this approach. In deleting a column, it should not be a target attribute or the primary attribute. It is not good to adopt if the dataset is very small.
- Replacing it with some aggregate value: If a few of the cell contain null values, we can assign some aggregate value, mean, median, or mode. This method can be adopted if the missing value is in the attribute of numeric values. This does not lead to loss of information as in the previous method.

- Prediction: The attribute having the missing values can be treated as a target variable. Thus, it can be predicted using the machine learning algorithm. It is crucial to remember while adopting this method, that the machine learning algorithm should not allow the presence of null values. Otherwise, the algorithm may learn null values as a valid value.

8.2 Data Preprocessing

The data collected from the sources is preprocessed by maintaining the file format and handling the missing values. So, the data contains all the accurate values, no missing value, and aligned attributes. Let's have a look at our dataset in Figure 8.3.

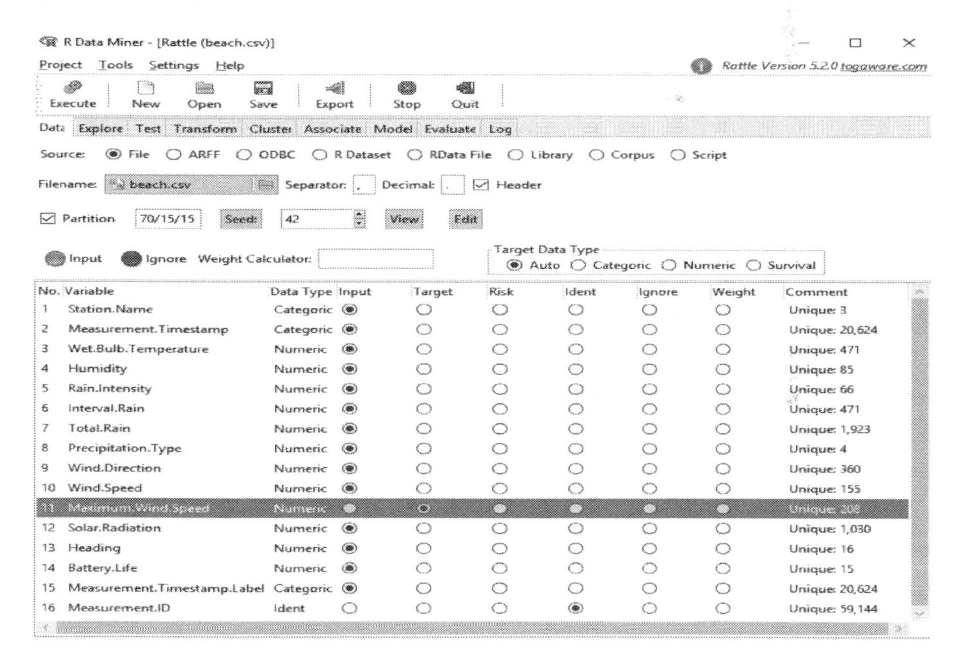

FIGURE 8.3: Preprocessed IoT experimental data in Rattle

There is only a limited number of options to view our data. In Rattle, we can view the summary of data. We know its impossible to check each and every instance. But, without checking, it's impossible to do data cleaning. Now the data is in the form of a complete file to be processed. But surprisingly, data is not clean yet. Cleaning means the data should not have outliers, no oversampling and no under-sampling. Let's learn about this in the next section.

8.3 Data Cleaning

Data cleaning refers to the phase of balancing the classes in the data. The imbalanced dataset may lead to inaccurate accuracies. What do we mean by imbalanced class or data? Suppose you are conducting a competition for primary classes in school. Due to the scheduled classes, the audience is limited. The last round of your competition is conducted with the participation of the audience. You may ask for volunteers to come on the stage and help the students. The volunteers from the audience who want to participate came on the stage. You observe 15 volunteers appeared, 13 volunteers among them belong to 6th grade or 7th grade but 2 belong to the 10th grade. Is it right to conduct the competition in this manner? It can be biased as the elder volunteers have more chances to win. The age factor should be considered important in such competitions. This is what we call imbalanced classes. There are many schemes available in the literature. These schemes give solutions with experiments. Let's briefly cover all the schemes.

8.3.1 The Condensed Nearest Neighbor Rule

This rule was developed by scientist Peter Hard in 1968. This rule is partially based on the nearest neighbor (NN) rule. It emphasis only on the attributes having imbalanced classes. The NN rule assigns the unclassified or misclassified instance in the class which is nearest to correctly classified instances. For point a, the rule searches for the nearest neighbor a$'$. If a and a$'$ have the same value or similar constraints, then they are placed in the same class. The NN rule starts from the center and moves to find the closest k labels.

The CNN rule works for selecting the subset of a particular set, by starting from a single instance. Then it compares the initial instance with the rest of the instances. The algorithm developed for the CNN rule is illustrated below.

Algorithm 1 The Condensed Nearest Neighbor Rule

Input: Ordered sample set in bin named S
Output: Classified instances.

```
1: procedure FUNCTION(PageRank)
2:     STORE= NULL                         ▷ Setting the store bin empty
3:     GRABBAG= NULL                       ▷ Setting the grabbag bin empty
4:     STORE= N_1                          ▷ Store the first instance of set N_1 in STORE bin
5:     while GRABBAG= NULL do
6:         for i = 1 to n do
7:             if S(n_i) == STORE(n_i) then
8:                 GRABBAG = (n_i)
9:             else
10:                STORE= (n_i)
11:            end if
12:        end for
13:    end while
14:    S(n) = STORE(n)
15: end procedure
```

Implementing CNN in R requires the understanding of its method and its argument. The syntax for the defining CNN is CNN (formula, data, ...). This function definition uses the arguments, the formula for describing the classification variable, and the attributes to be used, and data is training data used for experiments. The implementation of CNN in R is as shown in the algorithm and diagram that follows.

PROGRAM 1:

```
library(rpart)
library(rattle)
library(rpart.plot)
library(RColorBrewer)
library(party)
library(partykit)
library(caret)
library(NoiseFiltersR)
data(segmentationData)
segmentationData
data <- segmentationData[,-c(1,2)]
form <- as.formula(Class .)
out <- CNN(form, data)
print(out)
length(out$remIdx)
identical(out$cleanData, iris[setdiff(1:nrow(iris),out$remIdx),])
length(out$repIdx)
length(out$repLab)
```

OUTPUT:

```
> print(out)
Call:
CNN(formula = form, data = data)

Results:
Number of removed instances: 1119 (55.42348 %)
Number of repaired instances: 0 (0 %)
> length(out$remIdx)
[1] 1119
>identical(out$cleanData, segmentationData[setdiff(1:nrow
(segmentationData),out$remIdx),])
[1] FALSE
> length(out$repIdx)
[1] 0
> length(out$repLab)
[1] 0
```

The complete working is presented by Algorithm 1. To understand the functionality of CNN more clearly, the screenshot of CNN program in R is displayed in Figure 8.4. The complete program with code and output is shown above. CNN finds the consistent subset to correctly classify the remaining instances by the algorithm of NN. With the help of the first instance, it traverses through the complete set and finds the one which is completely classified.

Program 1 is implemented in R. Firstly, include the libraries required by CNN. We have retrieved some data using segmentation data function as done in previous examples. The output of CNN can be viewed by various parameters such as repIdx and repLab as discussed below.

- cleanData: A data frame which stores the filtered dataset.
- remIdx: This vector which contains the indexes of removed instances (vector of row numbers).
- repIdx: This vector contains the indexes of repaired instances.
- repLab: It presents the new labels of repaired instances.

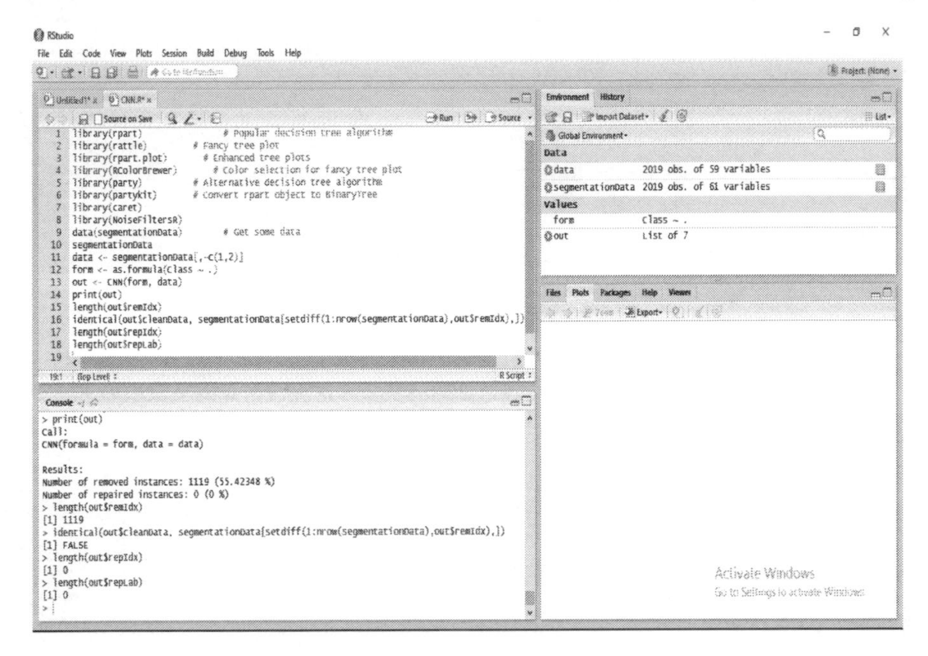

FIGURE 8.4: Implementation of condensed nearest neighbor rule in R

- parameters: It represents the argument values.
- call: It represents the function call.
- extraInf: It is extra interesting information.

8.3.2 Tomek

A function which uses Tomek links for imbalanced classification problems is named a TomekClassif function. It performs under-sampling by breaking one or more links, which in return breaks the Tomek link. The syntax for the TomekClassif function is as follows.

TomekClassif(form, data, dist = "Euclidean", Cl = "all", rem = "both")

In this function, the arguments used are discussed below:

- Form: It is the formula which describes the problem during the prediction.
- Data: It is the data which you are going to use for experimenting wmachine learning models. The dataset used may be imbalanced.
- Distance: There are various distance metrics to estimate the nearest neighbors. It is specified in characters. Some metric is defined, like for numeric instances – *Canberra, Manhattan, Chebyshev, Euclidean, p-norm,*

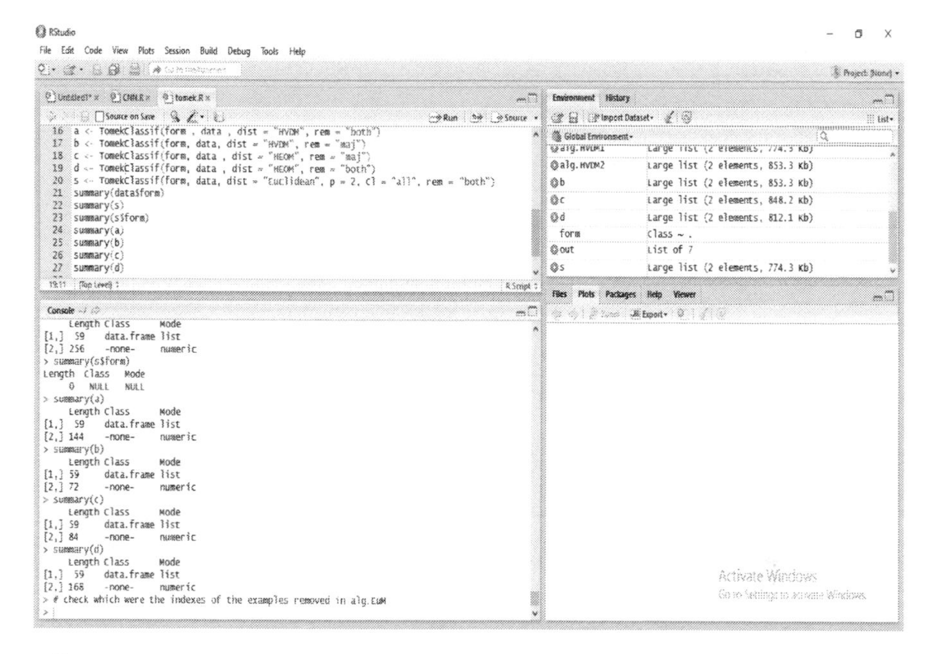

FIGURE 8.5: Implementation of Tomek links for imbalanced classification problem

for nominal instances its *overlap* and for both kind of instances, *HEOM* and *HVDM*.

- p: It is defined only in the case when the distance metric selected is p-norm. Its value is equivalent to p of p-norm.

- Cl: It defines the classes which are targeted to be under-sampled. It can be a specific subset of classes.

- rem: The Tomek links the two classes, if both the classes targeted are to be discarded, then it is set to *both*, otherwise for the major class its *maj*.

Program 2 illustrates the working of this TomekClassif function. The syntax of the functions and its parameters are defined previously. Figure 8.5 presents the implementation of this program in R.

PROGRAM 2:

```
library(ggplot2)
library(readr)
library(DMwR)
library(corrgram)
library(lattice)
library(grid)
library(unbalanced)
library(mlr)
library(ParamHelpers)
library(UBL)
data(segmentationData)
segmentationData
data <- segmentationData[,-c(1,2)]
form <- as.formula(Class   .)
a <- TomekClassif(form , data ,
dist = "HVDM", rem = "both")
b <- TomekClassif(form, data, dist
= "HVDM", rem = "maj")
c <- TomekClassif(form, data , dist
= "HEOM", rem = "maj")
d <- TomekClassif(form, data , dist
= "HEOM", rem = "both")
s <- TomekClassif(form, data, dist
= "Euclidean", p = 2, Cl = "all",
rem = "both")
summary(data$form)
summary(s)
summary(s$form)
summary(a)
summary(b)
summary(c)
summary(d)
```

OUTPUT:

```
> summary(data$form)
Length Class Mode
0 NULL NULL
> summary(s)
Length Class Mode
[1,] 59data.frame list
[2,] 256 -none- numeric
> summary(s$form)
Length Class Mode
0 NULL NULL
> summary(a)
Length Class Mode
[1,] 59data.frame list
[2,] 144 -none- numeric
> summary(b)
Length Class Mode
[1,] 59 data.frame list
[2,] 72 -none- numeric
> summary(c)
Length Class Mode
[1,] 59 data.frame list
[2,] 84 -none- numeric
> summary(d)
Length Class Mode
[1,] 59data.frame list
[2,] 168 -none- numeric
```

8.3.3 One-sided Selection

OSSClassif function is used by the one-sided selection strategy for handling multiclass.

Syntax of OSSClassif function is as follows.

OSSClassif(form, dat, dist = "Euclidean", p = 2, Cl = "smaller", start = "CNN")

The parameters used by the function are discussed as follows.

1. Form: It is the formula which describes the problem occurring in prediction.

2. Data: It is the data which can be used to experiment with the machine learning models. The dataset used may be imbalanced.

3. Distance: There are various distance metrics for estimating the nearest neighbors. It is specified in characters. Some metrics are defined like for numeric instances — *Canberra, Manhattan, Chebyshev, Euclidean, p-norm*; for nominal instances its *overlap* and for both kind of instances, *HEOM* and *HVDM*.

4. p: It is defined only in the case when the distance metric selected is p-norm. Its value is equivalent to *p* of p-norm.

5. Cl: It defines the classes which are targeted to be under-sampled. It can be the specific subset of classes.

6. Start: It determines the initial functionality, i.e., the strategy for the function. It can have two values: either CNN or Tomek. The default strategy is CNN. Both the strategies are applied, which can set the priority.

Program 3:
```
library(DMwR)
library(UBL)
data(segmentationData)
segmentationData
data <- segmentationData[,-c(1,2)]
form <- as.formula(Class .)
a <- OSSClassif(form, data, dist = "HVDM")
b <- OSSClassif(form, data, dist = "HEOM", start = "Tomek")
c <- OSSClassif(form, data, dist = "HVDM", start = "CNN")
d <- OSSClassif(form, data, dist = "HVDM", start = "Tomek")
e <- OSSClassif(form, data, dist = "HEOM")
summary(a)
summary(b)
summary(c)
summary(d)
summary(e)
```

OUTPUT:
Exercise for you

8.3.4 SMOTE

The SMOTE function handles imbalanced classification problems as illustrated in Figure 8.6. After implementing the variation in the SMOTE'd data, the output is presented in Figure 8.7. The syntax for this function is as follows:
SMOTE(form, data, perc.over = 200, k = 5, perc.under = 200, learner = NULL, ...)

The different parameters used in this function are discussed below.

1. form A: formula describing the working of the designed problem.
2. data: It is the unbalanced dataset, which is targeted to be balanced by this function.
3. perc.over: It is the argument which states the number of cases to be created for the minority cases. Generally, it is equal to 100 to maintain the ratio.
4. k: perc.over: This creates the number of cases for the minority number, but this argument indicates the frequency of nearest neighbors to be used for the generation of the instances.
5. perc.under: It is the argument which states the number of extra cases in the majority class to be selected for the cases of the minority class.

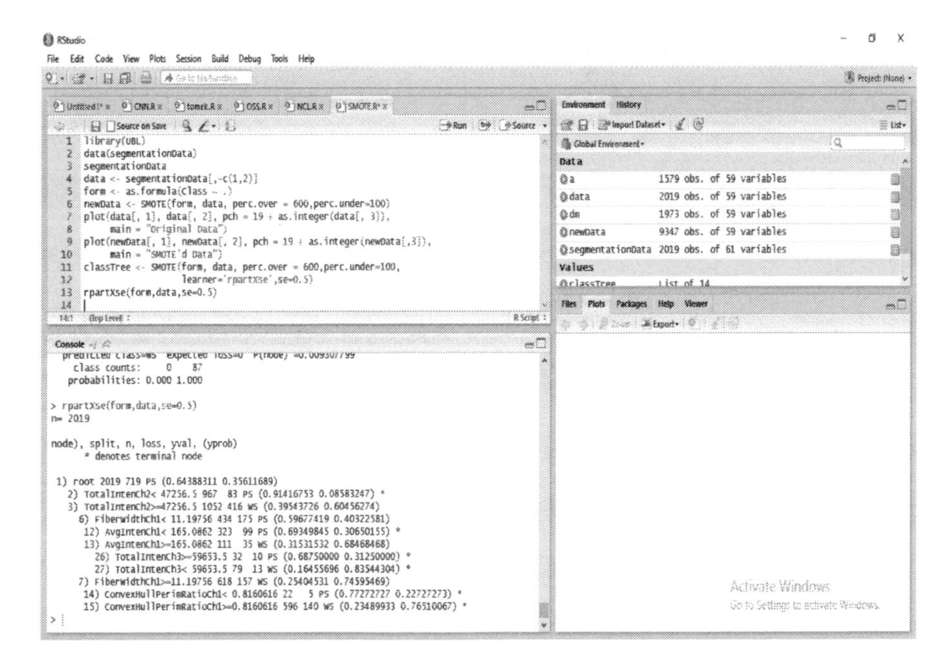

FIGURE 8.6: Implementation of SMOTE for unbalanced classification problems

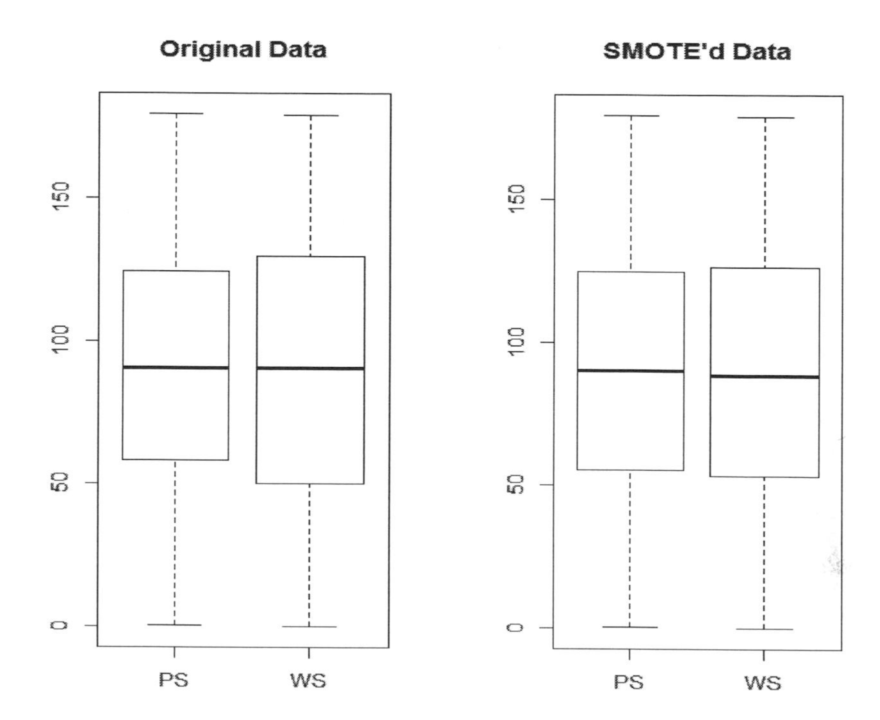

FIGURE 8.7: Plotting results of SMOTE for unbalanced classification problems

6. learner: It is an optional parameter used to specify the classification algorithm.

7. ...: In case you specify a learner (parameter learner), you can indicate f arguments which can be used when calling this learner.

PROGRAM 4:
```
library(UBL)
data(segmentationData)
segmentationData
data <- segmentationData[,-c(1,2)]
form <- as.formula(Class .)
newData<- SMOTE(form, data, perc.over = 600,perc.under=100)
plot(data[, 1], data[, 2], pch = 19 + as.integer(data[, 3]), main = "Original Data")
plot(newData[, 1], newData[, 2], pch = 19 + as.integer(newData[,3]), main = "SMOTE'd Data")
classTree<- SMOTE(form, data, perc.over = 600,perc.under=100, learner='rpartXse',se=0.5)
rpartXse(form,data,se=0.5)
```

OUTPUT: n= 2019 (number of instances in data)
```
1) root 2019 719 PS (0.64388311 0.35611689)
2) TotalIntenCh2< 47256.5 967 83 PS (0.91416753 0.08583247) *
3) TotalIntenCh2>=47256.5 1052 416 WS (0.39543726 0.60456274)
6) FiberWidthCh1< 11.19756 434 175 PS (0.59677419 0.40322581)
12) AvgIntenCh1< 165.0862 323 99 PS (0.69349845 0.30650155) *
13) AvgIntenCh1>=165.0862 111 35 WS (0.31531532 0.68468468)
26) TotalIntenCh3>=59653.5 32 10 PS (0.68750000 0.31250000) *
27) TotalIntenCh3< 59653.5 79 13 WS (0.16455696 0.83544304) *
7) FiberWidthCh1>=11.19756 618 157 WS (0.25404531 0.74595469)
14) ConvexHullPerimRatioCh1< 0.8160616 22 5 PS (0.77272727 0.22727273) *
15) ConvexHullPerimRatioCh1>=0.8160616 596 140 WS (0.23489933 0.76510067) *
```

8.3.5 ADASYN Algorithm

This function AdasynClassif provides the solution for unbalanced classification problems. It also works by creating the synthetic examples as done in SMOTE. But the underlying difference among the two is that this algorithm evaluates the weights of the class according to constraints defined in learning. Hence, this algorithm is more harder to learn as more synthetic data is required. The implementation of this algorithm is presented in Figure 8.8. The results of this algorithm are presented in Figure 8.9 and Figure 8.10. Its syntax is as follows.

AdasynClassif(form, dat, baseClass = NULL, beta = 1, dth = 0.95, k = 5, dist = "Euclidean", p = 2)

1. Form: It is the formula which describes the problem occurring in prediction.

2. Data: It is the data which can be used to experiment with the machine learning models. The dataset used may be imbalanced.

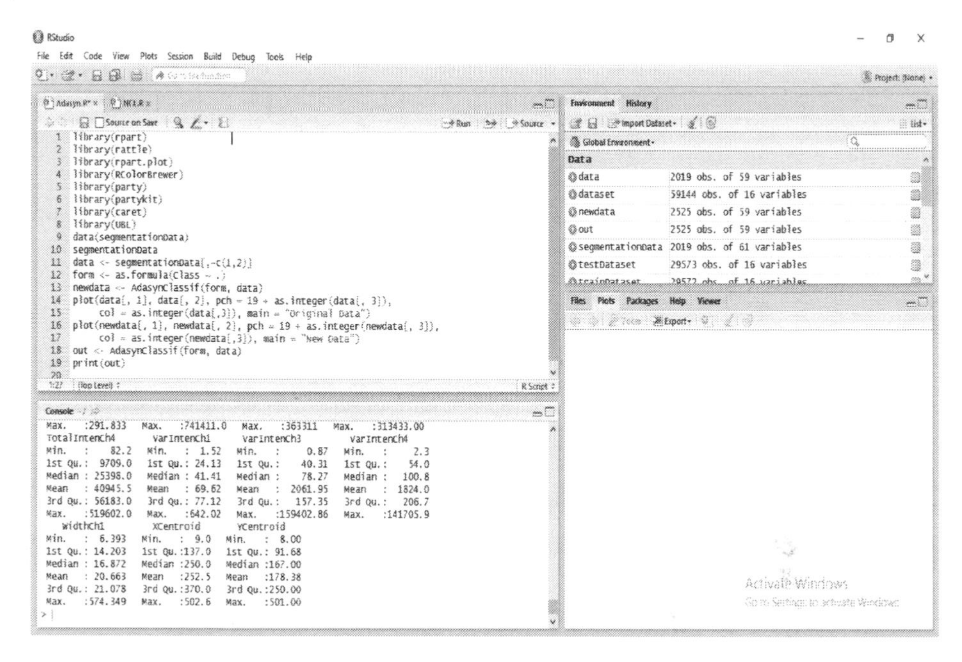

FIGURE 8.8: Implementation of ADASYN algorithm for unbalanced classification problems

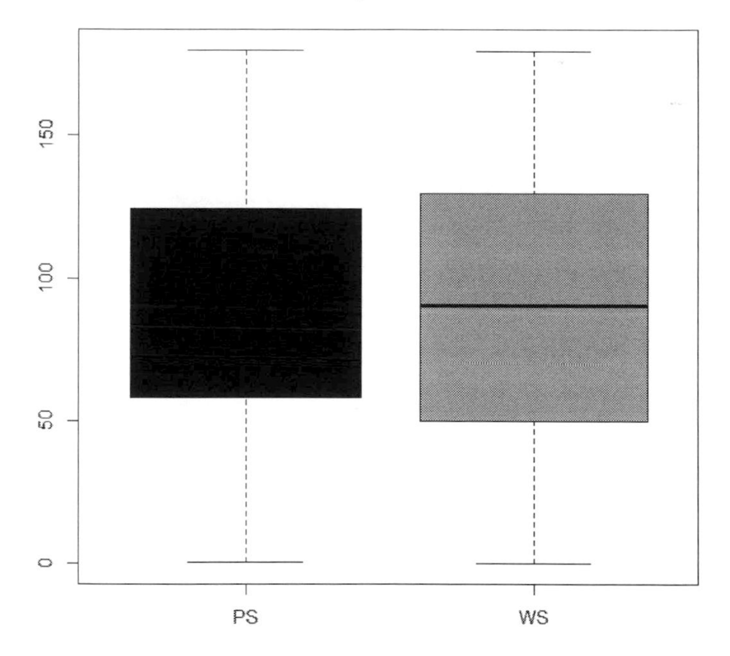

FIGURE 8.9: Data before applying ADASYN algorithm

3. Distance: There is various distance metrics for estimating the nearest neighbors. It is specified in characters. Some metrics are defined for numeric instances — *Canberra, Manhattan, Chebyshev, Euclidean, p-norm,* for nominal instances its *overlap*, and for both kind of instances, *HEOM* and *HVDM*.

4. p: It is defined only in the case when the distance metric selected is p-norm. Its value is equivalent to p of p-norm.

5. k: This argument indicates the frequency of the nearest neighbors to be used for the generation of the instances.

6. baseClass: It is the class which is used for comparison to the other classes. It considers the major class as default.

7. beta: It has either a numeric value indicating the desired balance level after synthetic examples generation or a named list specifying the selected classes, beta value. A beta value of 1 (the default) corresponds to full balancing the classes.

PROGRAM 5:

```
library(rpart)
library(rattle)
library(rpart.plot)
library(RColorBrewer)
library(party)
library(partykit)
library(caret)
library(UBL)
data(segmentationData)
segmentationData
data <- segmentationData[,-c(1,2)]
form <- as.formula(Class .)
newdata<- AdasynClassif(form,
data)
plot(data[, 1], data[, 2], pch =
19 + as.integer(data[, 3]), col =
as.integer(data[,3]), main = "Origi-
nal Data")
plot(newdata[, 1], newdata[, 2], pch
= 19 + as.integer(newdata[, 3]), col
= as.integer(newdata[,3]), main =
"New Data")
out <- AdasynClassif(form, data)
print(out)
```

Output:

```
Class
PS:1300
WS:1225
```

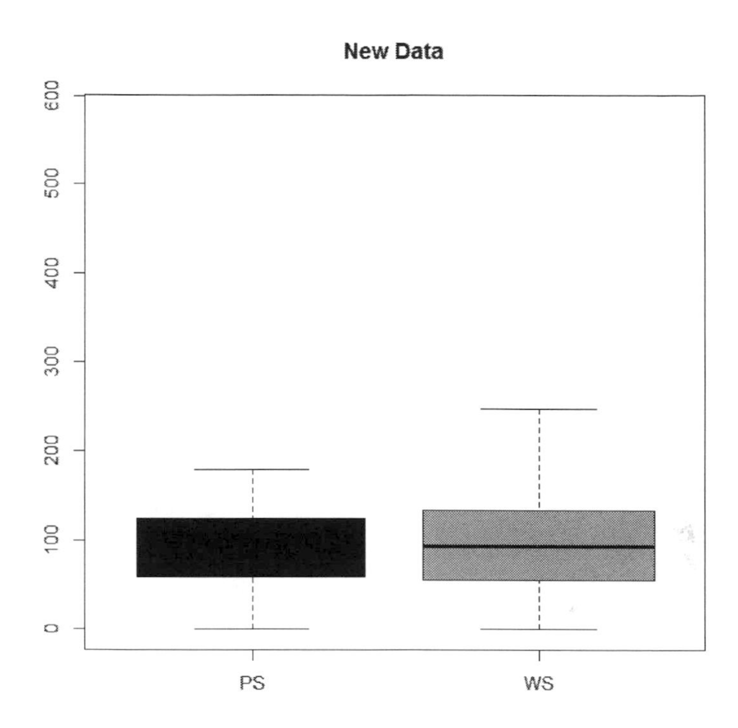

FIGURE 8.10: Data after applying ADASYN algorithm

8.3.6 SOTU

The idea behind this approach is to handle inaccurate accuracy due to imbalanced classes of data. For example, in the standard dataset, WEBSPAM-UK 2007, the spam labels are proportionally less in comparison to other labels as provided by the judges. It can lead to biasing of the classifiers. It is specially handled when the minor class, i.e, spam, is the target class for the experiments. The approach followed is *Split by Over-sampling and train by Under-fitting* (SOTU). It is the over-sampling method, by creating the multiple copies of minor class. The over-fitting is avoided by training the model by a single sample set at a time. Each set contains equal instances of the minor class and the major class (the spam class and the non-spam class with equal ratio). The complete dataset is divided into 11 buckets of sample sets. Each model is trained and tested 11 times with different sets. It has been therefore said that the system is trained with unbiased data. The sample distribution approach in the proposed data cleaning approach (SOTU) is depicted in Figure 8.11.

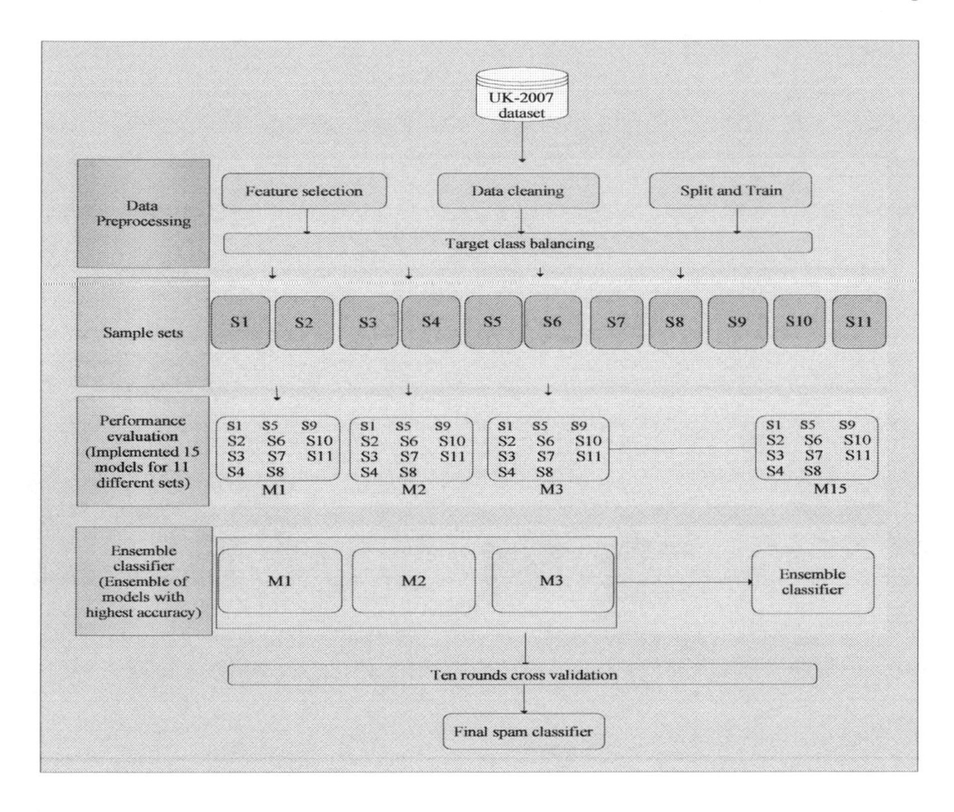

FIGURE 8.11: Architecture of 'split by over-sampling and train by underfitting' approach

$$S = \sum_{i=1}^{d} \frac{x_m + x_n}{d} \tag{8.1}$$

The Eq. 8.1 is used for the formulation of sample sets. S refers to the training file of the dataset. x_m refers to the instances of a minor class, and x_n refers to the instances of a major class. d is the number of sample sets; it depends upon the type of problem and number of features. In the case of a classification problem, it is fixed to be an odd number. In the case of a regression problem, it is fixed to be an even number. The number estimated should be close to the number of features (variation of 1 is suggested if needed). As a spam detection problem is a classification problem, so it is estimated to be an odd number. In this proposal, the feature count is 11, which is already an odd number. d is fixed with the value 11, and sample sets formed are 11.

8.4 Summary and What's Next?

The data to be processed in the machine learning model should be clean and noise-free. For this, various approaches used were elaborated on in this chapter. The data format should be consistent so different file formats are elaborated on in the chapter. Also, various data cleaning approaches are explained, such as condensed nearest neighbor rule, Tomek, one-sided selection, SMOTE, and many more as summarized in Table 8.2. These approaches are executed in R for better illustration.

TABLE 8.2: Different data cleaning approaches

No.	Approaches	Year	Technique	Description	Evaluation measures
1.	CNN [9]	1968	under-sampling	-	-
2.	Tomek links [33]	1976	under-sampling	-	-
3.	Balancing data [32]	1996	over sampling	created a new dataset with equal probabilities	error rate
4.	One-sided selection [18]	1997	under-sampling	under-sampling the majority class	geometric mean
5.	SHRINK system [17]	1998	best positive region	positive classification of an overlapping region of minority and majority classes	ROC
6.	Solution for data mining [20]	1998	over- and under-sampling	over-sampling of the minority class with under-sampling of the majority class	lift curve
7.	Learn to balance [13]	2000	focused re-sampling	introduced under-sampling, re-sampling, and a recognition-based induction scheme	error rate
8.	Neighborhood Cleaning Rule (NCL) [19]	2001	under-sampling	nearest neighbors that belong to the majority class are removed	accuracy
9.	Synthetic minority over-sampling technique (SMOTE) [5]	2002	over-sampling	the minority class is over-sampled by creating 'synthetic' examples	ROC
10.	SMOTE + Tomek links [3]	2003	over-sampling	Applied Tomek links to create class clusters more accurately	accuracy
11.	SOTU [23]	Proposed scheme	over-sampling	over-sampling but avoided over-fitting	accuracy

8.5 Exercises

Questions

1. What are the important constraints to be considered while data gathering and performing data aggregation?
2. What are the different file formats?
3. How should you handle the missing values in the data?
4. What is meant by data cleaning?
5. Explain the CNN method in detail.
6. Explain the Tomek approach in detail.
7. Explain the one-sided selection approach in detail.
8. Explain the approach followed by SMOTE.
9. Define the methodology followed by SOTU.
10. Briefly describe ADASYN algorithm with an example.

Multiple Choice Questions

1: How are the instances arranged in ARFF file?

(a) Ordered (b) Unordered

(c) Sorted (d) None of the above

2: What is the extension of the zipped file?

(a) .zip (b) .jar

(c) .exe (d) None of the above

3: The sparse file contains the maximum number of:

(a) Zeros (b) Ones

(c) Decimal values (d) None of the above

4: Which technique is adopted for handling missing data in the file?

(a) Deletion (b) Replacing

(c) Prediction (d) All of the above

5: Which parameter of CNN represents the labels of repaired instances?

(a) cleanData (b) repIdx

(c) repLab (d) remIdx

6: Which parameter of CNN is used to store the filtered data?

(a) cleanData (b) repIdx

(c) repLab (d) remIdx

7: What is the use of the 'p' parameter of Tomek function?

(a) p-norm (b) p-index

(c) p-data (d) None of the above

8: Which function is used by one-sided selection strategy?

(a) OSSClassif() (b) OSS()

(c) select() (d) None of the above

9: Which parameter is used to specify the classification algorithm in SMOTE?

(a) k (b) learner

(c) perc.over (d) perc.under

10: The 'Cl' parameter is used by data cleaning approaches to define the classes which are the target to be:

(a) Under-sampled
(b) Over-sampled
(c) Under-fitted
(d) Over-fitted

11: The 'baseClass' parameter is used by the ADASYN algorithm to define the class for:

(a) Comparison
(b) Prediction
(c) Target
(d) None of these

12: The SOTU approach uses the training class which is:

(a) Under-sampled
(b) Over-sampled
(c) Under-fitted
(d) Over-fitted

13: Which technique uses the approach of finding the best positive region?

(a) Tomek
(b) SHRINK
(c) SMOTE
(d) SOTU

9

Feature Engineering and Optimization

The machine learning algorithms work accurately with the appropriate instances and their attributes. We all know the instances are the real data world value, which is gathered from the real world itself. So the instance cannot be created, cannot be modified nor destroyed. The data is pre processed by instance engineering that we did in the last chapter. To make a machine learning model more reliable, it is important to select the appropriate attributes. In some algorithms, the attribute selection is done according to the working of the model. The attributes are similar to the constraints, but more attributes increase the complexity of the machine learning model. The results of the machine learning models highly depend upon the features.

It is important to select the appropriate attributes. As seen in the real datasets, the deeper you go, the more difficult it is to find appropriate attributes. This is because of the reason that the attributes are smaller in size. To deal with such an issue, it is necessary to know how well the attributes behave in the machine learning models. This is what we call feature engineering. Tree-based learning and decision-based learning are the most complex structures because they deal with the attributes. If we talk about instance-based learners, they are very impressionable. The decision trees treat the attributes in the divide and conquer approach, as the tree works by generating the condition-based formula for the attributes. In contrast, the naive Bayes algorithm treats every attribute independently, assuming that no relation resides among attributes. This is how the attributes affect the working of the machine learning model.

Sometimes, it is challenging to filter the relevant attributes. This is the case when the filtered attributes are less in number and strong dependency gets created. The target attribute then seems to adopt the values from the attribute with the same nominal range or learns exactly the similar patterns of the data. It may lead to inaccuracies. So, it is never advisable to select the attributes with some automatic/programmed attribute selection scheme. It is always best to understand the working of the machine learning model deeply and its agenda for the treatment of attributes. This helps determine the optimal set of features.

Another stage of handling attributes is dimensionality reduction. This does not reduce the attributes but reduces the dimension space. Sometimes, the dataset contains high-dimensional data which increases the complexity of the model. There are many dimension reduction schemes. The scheme followed for dimensionality reduction is commonly known as a feature extraction scheme. Let's explore both the concepts more deeply: feature extraction and feature selection.

9.1 Feature Reduction

Feature reduction is the approach used to reduce the high-dimensional features into low-dimensional features by forming the new feature space. However, feature extraction does not reduce the number of features but the complexity of the features.

9.1.1 Principal Component Analysis

Principal component analysis (PCA) is adopted for performing the task of feature extraction. PCA is a multivariate procedure that transforms the set of correlated variables into a set of uncorrelated variables. The new set is formed in such a way that the maximum variance is extracted from the variables. The extracted variance is removed, thus forming the second combination of variables. This process is continued until the correlation between the variables is diminished. It results in orthogonal (uncorrelated) variance and the procedure is known as the principal axis method. It reduces the dimension of the feature space. PCA is also known as a data compression technique. It reduces the complexity of data. The IoT data with numerical values have been considered for the experiments. To show the complete output, twelve features, namely, wet bulb temperature, humidity, rain intensity, interval rain, total rain, precipitation type, wind direction, wind speed, maximum wind speed, solar radiation, heading, battery life, and 500 rows are considered.

Principal component (PC) scores, which are also known as component scores, are computed. These are the scores computed by the multiplication of the standardized value of each row with each column. The computed PCs are the eigen values, which provide information about the variability in the data. There are different ways to perform PCA. The result is always a series of eigen values for PCs irrespective of the type of method used. We have implemented the proposed scheme in R. Different functions used for computing PCA are: prcomp(), primcomp(), and PCA(). These methods are described briefly:

1. prcomp() method: This function is built in the package called 'stats'. This method produces eigenvalues but does not compute PCs. The result after applying prcomp() generates a series of eigenvalues, but not the actual PCs. The syntax of this method is as follows.

prcomp(x, retx = TRUE, center = TRUE, scale. = FALSE, tol = NULL, rank. = NULL, ...)
The arguments acceptable by this function are:

- x: The data matrix, which is to be used for experiments while performing PCA.
- retx: It is an optional parameter which states whether the rotated variables should be returned.
- center: The variables are shifted towards zero or not.
- scale: It states whether the variables are to be scaled towards its variance or not.
- tol: A minimum level of components to be considered.
- rank: It is the number of principal components to be used

This function is used in Program 1. In this program, Lines 1 to 8 state the libraries required for the functionality of this program. Line 9 calls the read function with the object for reading the .csv file. Line 10 computes the PCA for the read data. Line 11 prints the standard deviation of the PCA computed. Line 12 returns the matrix of computed eigen values. Line 13 prints the values rotated by the retx argument. The output of this program is appended as below.

PROGRAM 1:

```
1. library(randomForest)
2. library(hmeasure)
3. library(FSelector)
4. library(FactoMineR)
5. library(ade4)
6. library(ggplot2)
7. library(FactoMineR)
8. library(devtools)
9. data<- read.csv("F:/experiments/beach1.csv")
10. pca1 = prcomp(data, scale. = TRUE)
11. pca1$sdev
12. head(pca1$rotation)
13. head(pca1$x)
```

OUTPUT:

```
>library(hmeasure)
>library(FSelector)
>library(FactoMineR)
>library(ade4)
>library(ggplot2)
> library(FactoMineR)
>library(devtools)
>data<- read.csv("F:/experiments/beach1.csv")
> pca1 = prcomp(data, scale. = TRUE)
> pca1$sdev
[1] 1.8354207 1.5827218 1.1949725 1.0378641 0.9563216 0.9015651
0.8540104 0.7168229
[9] 0.6568880 0.3806957 0.2128649 0.1697634
>head(pca1$rotation)
PC1 PC2 PC3 PC4 PC5
Wet.Bulb.Temperature -0.45845244 0.18215272 -0.092028237
0.172864649 -0.21873624
Humidity -0.06123625 -0.37469727 -0.001886231 -0.145942937
0.24788960
Rain.Intensity -0.12158475 -0.22506254 -0.531559993 -0.208895457
0.10955651
Interval.Rain -0.01426211 -0.29176049 -0.411138003 -0.128106431
-0.08623425
Total.Rain -0.49893019 0.01311483 0.117848444 0.018942402
-0.02234642
Precipitation.Type -0.20559674 -0.26233369 -0.459994477 -
0.003149514 -0.05708889
PC6 PC7 PC8 PC9 PC10
Wet.Bulb.Temperature 0.098088033 -0.008226882 0.1355208
-0.40218039 -0.35566929
Humidity 0.765751244 0.177364975 -0.3051986 0.01963550 -
0.24820352
Rain.Intensity -0.433590247 0.179869377 -0.5341351 -0.29953155
0.03021242
Interval.Rain 0.129258330 -0.820302111 0.1725736 0.02878371
0.03855111
Total.Rain 0.206135694 -0.009293782 -0.1030771 -0.07368064
0.82133309
Precipitation.Type -0.005997324 0.465931786 0.5567120
0.37908597 0.05498776
PC11 PC12
Wet.Bulb.Temperature -0.559418450 0.202956339
```

Continued:

```
Humidity -0.010739731 0.003818422
Rain.Intensity -0.007415427 -0.005228646
Interval.Rain 0.007804646 -0.005459867
Total.Rain -0.053193431 -0.010031050
Precipitation.Type 0.005113261 0.015309847
>head(pca1$x)
PC1 PC2 PC3 PC4 PC5 PC6 PC7
[1,] -0.7533908 1.3603800 0.60365316 -1.77415528 -1.35495390
-1.273898 -0.2977139
[2,] -0.7159550 1.2824627 0.20512505 -0.01673742 0.09258246
-1.275718 -0.4999055
[3,] -0.6957618 1.0312744 0.27934533 0.41413854 -0.08968140
-1.455666 -0.5261064
[4,] -0.7396331 0.9386153 0.33384239 0.57431221 0.01920235
-1.573593 -0.5571894
[5,] -0.7363774 1.4551409 0.03254238 0.62289063 0.34408267
-1.506005 -0.6206552
[6,] -0.7978502 1.2135378 0.01373142 0.73447240 0.33364430
-1.114326 -0.5286868
PC8 PC9 PC10 PC11 PC12
[1,] -0.033381035 0.9311134 -0.7819385 0.4349608 0.013254818
[2,] -0.065647058 0.9304586 -0.7023308 0.5242046 -0.007462584
[3,]-0.073486202 0.8762675 -0.6750731 0.5108724 -0.141074139
[4,] 0.012865583 0.8008849 -0.6835795 0.4666068 -0.569932200
[5,] -0.009268147 0.9288297 -0.6345083 0.5900327 -0.129571878
[6,] -0.201549093 0.9027729 -0.7955311 0.5251831 -0.084495022
```

2. Using the primcomp() method: It is similar to the prcomp() method and comes with the 'stats' package. The difference between the two is that the primcomp() gives the mirror image of values instead of actual computed values. The result after applying primcomp() produces a series of eigenvalues. The exact PCs are not computed by this method. The syntax of this function is as follows:
   princomp(x, cor = FALSE, scores = TRUE, covmat = NULL, subset = rep_len(TRUE, nrow(as.matrix(x))), fix_sign = TRUE, ...) The functions accept various arguments as discussed below.

   - x: The data matrix to be used for experiments while performing PCA.
   - cor: It specifies the matrix to be used for calculation, either the correlation or the covariance matrix.
   - scores: It computes the score of each principal component.

- covmat: This is a covariance matrix to be used instead of the covariance matrix of x.
- fix_sign: It is the sign of the scores to be chosen.

We have used this function in program 2. Line 1 in this program returns the computed PCs in variable pca2. Line 2 computes the standard deviations of the value in pca2. Line 3 prints the matrix of computed eigenvalues. Line 4 prints the scores of each principal component. The output of this program is appended at the end of the program.

PROGRAM 2:

```
1. pca2 = princomp(data, cor = TRUE)
2. pca2$sdev
3. unclass(pca2$loadings)
4. head(pca2$scores)
```

OUTPUT:

```
> pca2 = princomp(data, cor = TRUE)
> pca2$sdev
Comp.1 Comp.2 Comp.3 Comp.4 Comp.5 Comp.6
Comp.7 Comp.8 Comp.9
1.8354207 1.5827218 1.1949725 1.0378641 0.9563216 0.9015651
0.8540104 0.7168229 0.6568880
Comp.10 Comp.11 Comp.12
0.3806957 0.2128649 0.1697634
>unclass(pca2$loadings)
Comp.1 Comp.2 Comp.3 Comp.4 Comp.5
Wet.Bulb.Temperature 0.45845244 0.18215272 0.092028237
0.172864649 0.21873624
Humidity 0.06123625 -0.37469727 0.001886231 -0.145942937
-0.24788960
Rain.Intensity 0.12158475 -0.22506254 0.531559993 -0.208895457
-0.10955651
Interval.Rain 0.01426211 -0.29176049 0.411138003 -0.128106431
0.08623425
Total.Rain 0.49893019 0.01311483 -0.117848444 0.018942402
0.02234642
Precipitation.Type 0.20559674 -0.26233369 0.459994477 -
0.003149514 0.05708889
```

Continued:

Wind.Direction -0.16120491 0.04522763 0.282489389 0.650800751 0.48761543
Wind.Speed -0.06288441 -0.55358689 -0.312656986 0.091902728 0.17264949
Maximum.Wind.Speed 0.06467973 -0.53704074 -0.297924488 0.159114244 0.25685763
Solar.Radiation -0.05655801 0.10858394 -0.060526774 -0.654214287 0.71576282
Heading 0.41499927 -0.05101608 -0.207012980 -0.030569154 -0.09678055
Battery.Life -0.51993819 -0.10170644 0.040275007 -0.046532681 -0.11078058
Comp.6 Comp.7 Comp.8 Comp.9 Comp.10
Wet.Bulb.Temperature 0.098088033 0.008226882 0.13552082 0.40218039 0.35566929
Humidity 0.765751244 -0.177364975 -0.30519856 -0.01963550 0.24820352
Rain.Intensity -0.433590247 -0.179869377 -0.53413507 0.29953155 -0.03021242
Interval.Rain 0.129258330 0.820302111 0.17257365 -0.02878371 -0.03855111
Total.Rain 0.206135694 0.009293782 -0.10307712 0.07368064 -0.82133309
Precipitation.Type -0.005997324 -0.465931786 0.55671202 -0.37908597 -0.05498776
Wind.Direction 0.167930285 -0.029497147 -0.37956274 -0.22359884 -0.04504366
Wind.Speed -0.169124892 -0.017036608 0.02781177 0.09599852 -0.04715204
Maximum.Wind.Speed -0.150254500 -0.029051841 0.06541583 0.17816952 0.06882630
Solar.Radiation 0.092903575 -0.102510824 -0.10698115 -0.09992405 0.00828277
Heading -0.262174067 0.176465464 -0.30291047 -0.69463776 0.20452068
Battery.Life 0.004595700 -0.048926145 -0.02307648 -0.12132500 -0.28459985
Comp.11 Comp.12
Wet.Bulb.Temperature 0.559418450 0.202956339
Humidity 0.010739731 0.003818422
Rain.Intensity 0.007415427 -0.005228646
Interval.Rain -0.007804646 -0.005459867

Continued:

```
Total.Rain 0.053193431 -0.010031050
Precipitation.Type -0.005113261 0.015309847
Wind.Direction -0.073952111 0.014964313
Wind.Speed -0.086372958 0.710642293
Maximum.Wind.Speed 0.104336366 -0.672520949
Solar.Radiation 0.019523299 0.003923918
Heading 0.235591991 0.023101387
Battery.Life 0.777344225 0.018223204
>head(pca2$scores)
Comp.1 Comp.2 Comp.3 Comp.4 Comp.5 Comp.6 Comp.7
[1,] 0.7541454 1.3617424 -0.60425772 -1.77593210 1.35631089
-1.275174 0.2980121
[2,] 0.7166720 1.2837471 -0.20533049 -0.01675418 -0.09267518
-1.276996 0.5004062
[3,] 0.6964586 1.0323072 -0.27962509 0.41455330 0.08977122
-1.457124 0.5266333
[4,] 0.7403739 0.9395554 -0.33417673 0.57488739 -0.01922158
-1.575169 0.5577474
[5,] 0.7371149 1.4565982 -0.03257497 0.62351446 -0.34442727
-1.507513 0.6212768
[6,] 0.7986492 1.2147532 -0.01374517 0.73520798 -0.33397845
-1.115442 0.5292163
Comp.8 Comp.9 Comp.10 Comp.11 Comp.12
[1,] -0.033414466 -0.9320459 0.7827216 -0.4353964 0.013268092
[2,] -0.065712804 -0.9313905 0.7030341 -0.5247296 -0.007470058
[3,] -0.073559799 -0.8771451 0.6757492 -0.5113841 -0.141215425
[4,] 0.012878468 -0.8016870 0.6842641 -0.4670741 -0.570502989
[5,] -0.009277429 -0.9297599 0.6351438 -0.5906236 -0.129701644
[6,] -0.201750945 -0.9036770 0.7963279 -0.5257090 -0.084579644
```

3. pca() method: It is one of the best ways to perform PCA. This method is included in 'FactomineR' package. It produces the exact series of PC corresponding to each row for each column. The pca() works in such a way that it reduces the variance among the features. The syntax of this function is:
   PCA(X, scale.unit = TRUE, ncp = 5, ind.sup = NULL, quanti.sup = NULL, quali.sup = NULL, row.w = NULL, col.w = NULL, graph = TRUE, axes = c(1,2))
   The arguments accepted by this function are:

   - x: The data matrix to be used for experiments while performing PCA.

- ncp: The number of dimensions to be kept in the results which are 5 by default.
- scale.unit: It is true if the data is scaled to the variance.
- ind.sup: It specifies the indexes of supplementary instances.
- quanti.sup: It specifies the indexes of quantitative supplementary instances.
- quali.sup: It specifies the indexes of categorical supplementary instances.
- row.w: It specifies the weights of rows.
- col.w: It specifies the weights of columns.
- graph: If it is TRUE, then a graph is displayed.
- axes: It is the vector specifying the components to the plot.

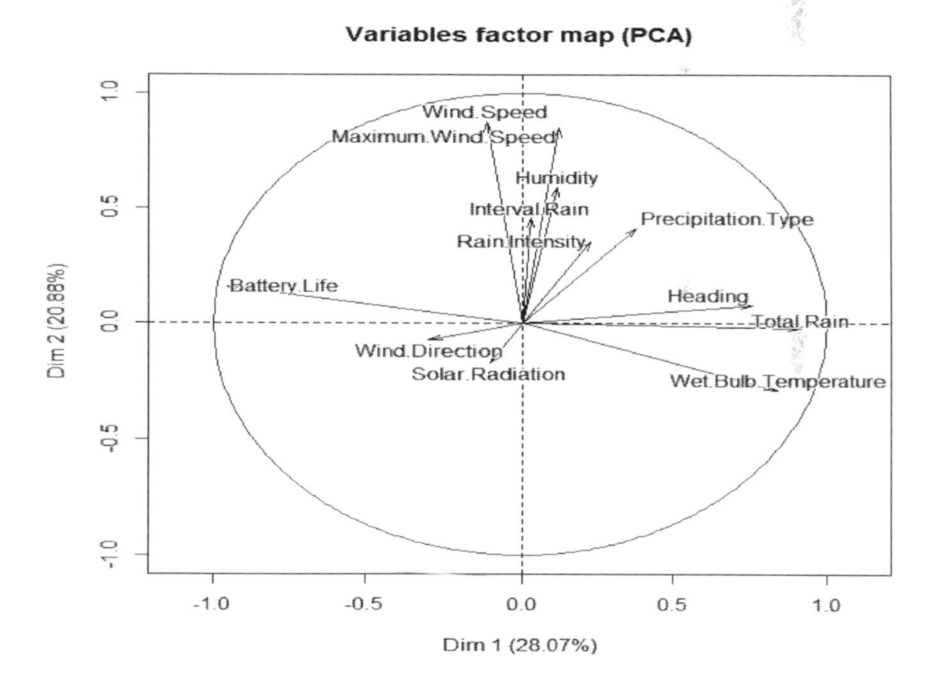

FIGURE 9.1: PCA using pca() method

This function is used in program 3. Line 1 draws Figure 9.1. Line 2 returns the eigenvalues of the computed PCs. The matrix of active variables is returned by Line 3. Line 4 returns the matrix of active individuals.

PROGRAM 3:

```
1. pca3 = PCA(data, graph = TRUE)
2. pca3$eig
3. pca3$var$coord
4. head(pca3$ind$coord)
```

OUTPUT:

```
> pca3$eig
eigenvalue percentage of variance cumulative percentage of variance
comp 1 3.36876913 28.0730761 28.07308
comp 2 2.50500828 20.8750690 48.94815
comp 3 1.42795928 11.8996607 60.84781
comp 4 1.07716183 8.9763486 69.82415
comp 5 0.91455095 7.6212579 77.44541
comp 6 0.81281966 6.7734972 84.21891
comp 7 0.72933373 6.0777811 90.29669
comp 8 0.51383511 4.2819592 94.57865
comp 9 0.43150179 3.5958482 98.17450
comp 10 0.14492920 1.2077433 99.38224
comp 11 0.04531145 0.3775954 99.75984
comp 12 0.02881960 0.2401633 100.00000
> pca3$var$coord
Dim.1 Dim.2 Dim.3 Dim.4 Dim.5
Wet.Bulb.Temperature 0.84145309 -0.28829708 0.109971212 -
0.179410008 0.20918219
Humidity 0.11239428 0.59304154 0.002253994 0.151468931 -0.23706218
Rain.Intensity 0.22315917 0.35621138 0.635199575 0.216805089 -
0.10477126
Interval.Rain 0.02617697 0.46177569 0.491298609 0.132957062
0.08246768
Total.Rain 0.91574680 -0.02075713 -0.140825650 -0.019659638
0.02137036
Precipitation.Type 0.37735652 0.41520124 0.549680752 0.003268767
0.05459533
Wind.Direction -0.29587882 -0.07158275 0.337567052 -0.675442717
0.46631716
Wind.Speed -0.11541934 0.87617404 -0.373616501 -0.095382539
0.16510843
Maximum.Wind.Speed 0.11871451 0.84998608 -0.356011571 -
0.165138957 0.24563849
```

Continued:

```
Solar.Radiation -0.10380774 -0.17185816 -0.072327830 0.678985503
0.68449942
Heading 0.76169825 0.08074427 -0.247374819 0.031726627 -0.09255333
Battery.Life -0.95430532 0.16097301 0.048127526 0.048294598 -
0.10594185
>head(pca3$ind$coord)
1 0.7541454 -1.3617424 -0.60425772 1.77593210 1.35631089
2 0.7166720 -1.2837471 -0.20533049 0.01675418 -0.09267518
3 0.6964586 -1.0323072 -0.27962509 -0.41455330 0.08977122
4 0.7403739 -0.9395554 -0.33417673 -0.57488739 -0.01922158
5 0.7371149 -1.4565982 -0.03257497 -0.62351446 -0.34442727
6 0.7986492 -1.2147532 -0.01374517 -0.73520798 -0.33397845
```

PROGRAM 4:

```
1. g<- ggbiplot(newde, obs.scale = 1, var.scale = 1, ellipse = TRUE,
circle = TRUE)
g <- g + scale_color_discrete(name = ")
g <- g + theme(legend.direction = 'horizontal', legend.position =
'top')
2. print(g)
3. circle<- function(center = c(0, 0), npoints = 100)
r = 1
tt = seq(0, 2 * pi, length = npoints)
xx = center[1] + r * cos(tt)
yy = center[1] + r * sin(tt)
return(data.frame(x = xx, y = yy))
4. corcir = circle(c(0, 0), npoints = 100)
5. correlations = as.data.frame(cor(data, pca1$x))
6. arrows = data.frame(x1 = c(0, 0, 0, 0), y1 = c(0, 0, 0, 0), x2 =
correlations$PC1, y2 = correlations$PC2)
7. ggplot() + geom_path(data = corcir, aes(x = x, y = y), colour =
"gray65") +
geom_segment(data = arrows, aes(x = x1, y = y1, xend = x2, yend
= y2), colour = "gray65") +
geom_text(data = correlations, aes(x = PC1, y = PC2, label =
rownames(correlations))) +
geom_hline(yintercept = 0, colour = "gray65") +
geom_vline(xintercept = 0,
colour = "gray65") + xlim(-1.1, 1.1) + ylim(-1.1, 1.1) + labs(x =
"pc1 aixs",
y = "pc2 axis") + ggtitle("Circle of correlations")
```

OUTPUT:

```
>circle<- function(center = c(0, 0), npoints = 100)
+ r = 1
+ tt = seq(0, 2 * pi, length = npoints)
+ xx = center[1] + r * cos(tt)
+ yy = center[1] + r * sin(tt)
+ return(data.frame(x = xx, y = yy))
+
>corcir = circle(c(0, 0), npoints = 100)
>correlations = as.data.frame(cor(data, pca1$x))
> # data frame with arrows coordinates
>arrows = data.frame(x1 = c(0, 0, 0, 0), y1 = c(0, 0, 0, 0), x2 =
correlations$PC1,
+ y2 = correlations$PC2)
> # geom_path will do open circles
>ggplot() + geom_path(data = corcir, aes(x = x, y = y), colour =
"gray65") +
+ geom_segment(data = arrows, aes(x = x1, y = y1, xend = x2,
yend = y2), colour = "gray65") +
+ geom_text(data = correlations, aes(x = PC1, y = PC2, label =
rownames(correlations))) +
+ geom_hline(yintercept = 0, colour = "gray65") +
geom_vline(xintercept = 0,
+ colour = "gray65") + xlim(-1.1, 1.1) + ylim(-1.1, 1.1) + labs(x =
"pc1 aixs",
+ y = "pc2 axis") + ggtitle("Circle of correlations")
```

Program 4 prints the required graphs of PCA. Line 1 returns the variance between two PCs into the variable g. Line 2 prints the variance stored in variable g in Figure 9.2. Program 4 find the correlation among the attributes and draw the circle of correlation in Figure 9.3.

9.2 Feature Selection

Feature selection aims at removing the irrelevant and redundant features by considering a small set of features. The set of relevant features targets to improve the performance of the machine learning model. The feature selection process has various advantages, such as data reduction. Knowledge extraction from the process. Limiting the storage requirements; reducing the cost in terms of time and space complexities; and data understanding. The num-

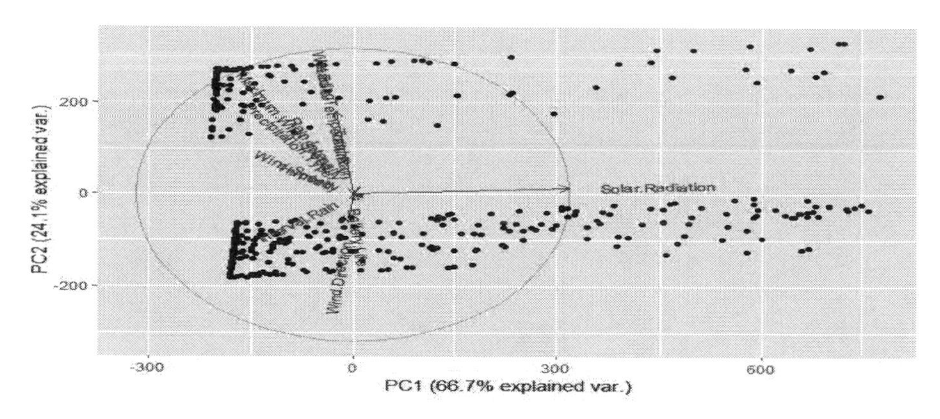

FIGURE 9.2: Variance between two PCs

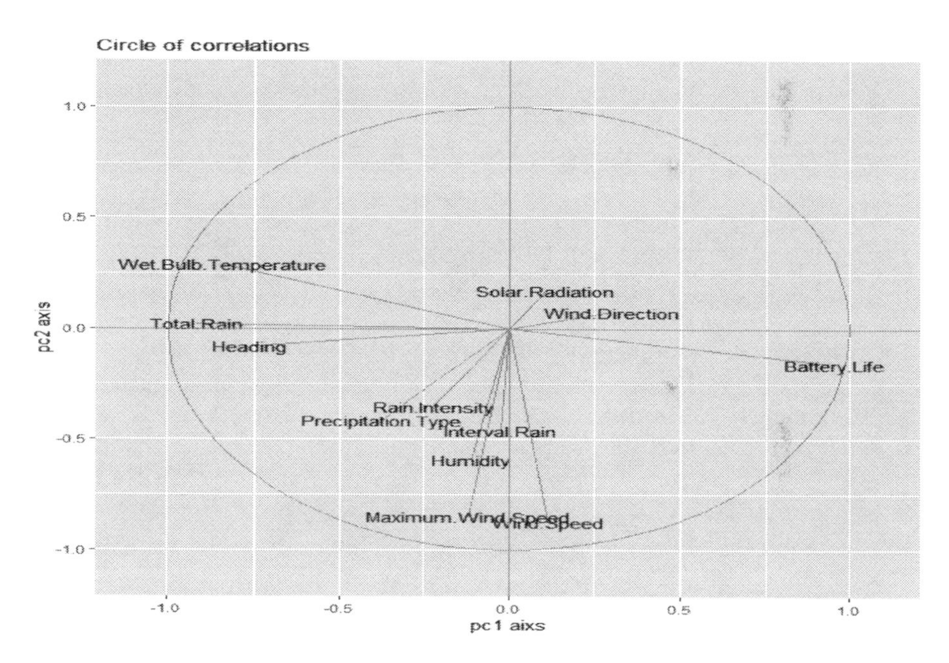

FIGURE 9.3: ggplot function draws the relation among different attributes

ber of feature selection techniques exists in the literature. The two different categories of feature selection techniques are formed and discussed as below.

1. Filter method: These methods consider the general characteristics of features and ignore the relevancy of predictor. These methods are less expensive. The filter methods are independent of the machine learning model. The list of filter methods with their arguments is shown in Table 9.1. All of these methods require the FSelector library for inclusion in the program.

TABLE 9.1: Filter methods

Method	**Methodology**	**Arguments**	**Values**
relief	works by finding the distance between instances	formula, data, neighbours.count, sample.size	data.frame
best.first.search	works by searching attribute subset space	attributes, eval.fun, max.backtracks	vector
cfs	works by finding correlation	formula, data	vector
chi.squared	works by performing chi-square test	formula, data	data.frame
consistency	works by consistency measure	formula, data	vector
correlation	finds weights of continuous attributes by computing correlation	formula, data	data.frame
information.gain	finds weights of discrete attributes by computing correlation	formula, data, unit	data.frame
greedy.search	implements greedy search algorithm	attributes, eval.fun	vector
hill.climbing. search	works by finding the local maximum	attributes, eval.fun	vector
oneR	applies the association rules	formula, data	data.frame
random.forest. importance	implements RandomForest algorithm	formula, data, importance.type	

2. Wrapper method: These methods consider the predictor as a part of their working. These methods are more expensive but are more accurate. The three commonly used wrapper methods are as shown in Table 9.2.

9.2.1 Feature Importance

There are numerous filter methods in the literature. These methods find the importance of the features and treat each feature independently. Some of these methods are cfs(), chi-squared(), information.gain(), gain.ratio(), symmetrical.uncertainty(), linear.correlation(), rank.correlation(), oneR, relief, consis-

TABLE 9.2: Wrapper methods

Method	Methodology	Arguments	Values
Forward	It implements forward selection algorithm	X_train, y_train, X_val, y_val, min_change, n_features, criterion, verbose	Vector
Backward	It implements backward selection algorithm	X_train, y_train, X_val, y_val, min_change, n_features, criterion, verbose	Vector
Recursive Feature Elimination	It works by recursively eliminating the features.	x, y, sizes, rfecontrol	Vector

tency, random.forest.importance(). All of these functions make use of the FSelector package. These methods are elaborated after the exceptional function which uses the caret package. We have implemented the randomForest function in Program 5. This program returns(to) us the importance of each feature. This importance score is shown in Figure 9.4.

Program 5:

```
library(caret)
library(randomForest)
data<- read.csv("F:/beach1.csv")
Feature_importance=randomForest(data)
varImpPlot(Feature_importance)
```

9.2.1.1 Chi-squared Filter

The chi-square test is performed by this filter to find the weights of discrete attributes. The complete working of this function is elaborated in Program 6. The results are presented in Table 9.3. The syntax of this function is as follows.
chi.squared(formula, data)
The arguments used in the function definition are described as follows.

1. formula: It is the description of the working behind the algorithm.
2. data: It is the set of training data with the defined attributes for which the selection is to be made.

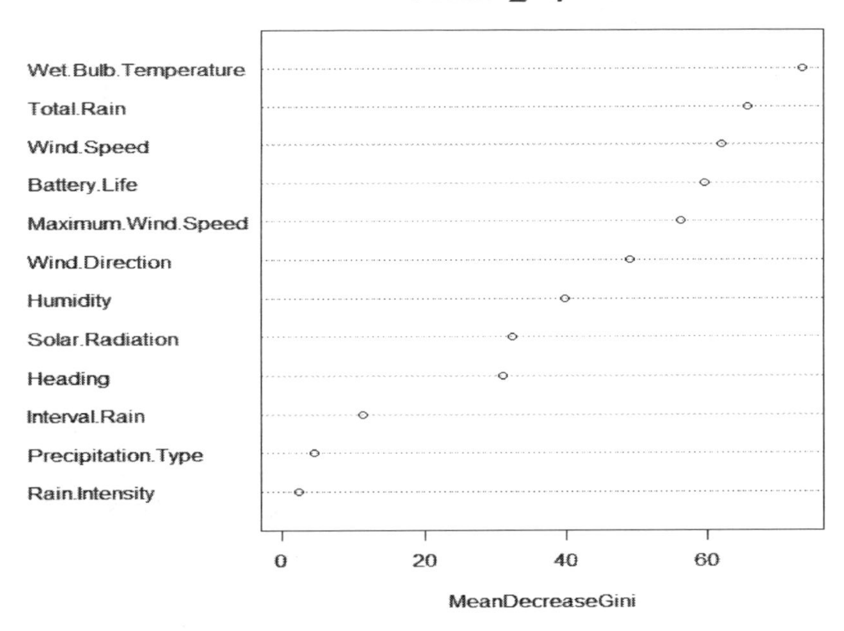

FIGURE 9.4: The importance of independent features

PROGRAM 6:

```
library(FSelector)
cat("Step 1: Enter the data")
InputDataFileName="F:/beach.csv"
InputDataFileName
dataset<- read.csv(InputDataFileName) # Read the datafile
dataset<- dataset[sample(nrow(dataset)),] # Shuffle the data row
wise.
cat("Step 2: Choose Target Variable")
target <- names(dataset)[1] # i.e. "Wet.Bulb.Temperature"
target
cat("Step 3: Create formula: symbolic creation of data -> ", target)
formula<- as.formula(paste(target, " ", paste(c(inputs), collapse =
"+")))
formula
b= chi.squared(formula, data) # Chi-squared filter
b
```

TABLE 9.3: Results of chi-squared filter

Feature	attr_importance
Humidity	0.3244248
Rain.Intensity	0.0000000
Total.Rain	0.6176635
Precipitation.Type	0.2158142
Wind.Direction	0.4464516
Wind.Speed	0.4522867
Maximum.Wind.Speed	0.2359112
Solar.Radiation	0.0000000
Heading	0.5070832
Battery.Life	0.7002321

9.2.1.2 Consistency-based Filter

This filter uses the consistency measure for computing the best subset of discrete and continuous attributes. The algorithm in the working of this filter is best.first.search. For illustrating the concept used in this filter, Program 7 is designed. The syntax of this function is:
consistency(formula, data)

The arguments used in the function definition are described here.

1. formula: It is the description of the working behind the algorithm.
2. data: It is the set of training data with the defined attributes for which the selection is to be made.

PROGRAM 7:

```
library(FSelector)
cat("Step 1: Enter the data")
InputDataFileName="F:/beach.csv"
InputDataFileName
dataset<- read.csv(InputDataFileName) # Read the datafile
dataset<- dataset[sample(nrow(dataset)),] # Shuffle the data row wise.
cat("Step 2: Choose Target Variable")
target <- names(dataset)[1] # i.e. "Wet.Bulb.Temperature"
target
cat("Step 3: Create formula: symbolic creation of data -> ", target)
formula<- as.formula(paste(target, " ", paste(c(inputs), collapse =
"+")))
formula
c= consistency(formula, data) # Consistency-based filter
c
```

OUTPUT:

```
[1] "Humidity" "Total.Rain" "Wind.Direction" "Wind.Speed"
[5] "Maximum.Wind.Speed" "Heading" "Battery.Life"
```

9.2.1.3 Correlation Filter

This algorithm computes the correlation among the continuous attributes to find their weights. Two functions are defined for using this filter. The first is linear.correlation (Program 8) which uses Pearson's correlation for its working and whose results are presented in Table 9.4. The second is rank.correlation (Program 9), which uses Spearman's correlation for its computation, and the results are presented in Table 9.5. The syntax for these functions is:
linear.correlation(formula, data)
rank.correlation(formula, data)
The arguments used in the function definition are described here.

1. formula: It is the description and working of the algorithm.
2. data: It is the set of training data with the defined attributes for which the selection is to be made.

PROGRAM 8:

```
library(FSelector)
cat("Step 1: Enter the data")
InputDataFileName="F:/beach.csv"
InputDataFileName
dataset<- read.csv(InputDataFileName) # Read the datafile
dataset<- dataset[sample(nrow(dataset)),] # Shuffle the data row
wise.
cat("Step 2: Choose Target Variable")
target <- names(dataset)[1] # i.e. "Wet.Bulb.Temperature"
target
cat("Step 3: Create formula: symbolic creation of data -> ", target)
formula<- as.formula(paste(target, " ", paste(c(inputs), collapse =
"+")))
formula
d= linear.correlation(formula, data) # Correlation filter
d
```

TABLE 9.4: Results of correlation filter

Feature	**attr_importance**
Humidity	0.10449798
Rain.Intensity	0.07186169
Interval.Rain	0.04364372
Total.Rain	0.75010476
Precipitation.Type	0.23594083
Wind.Direction	0.02861378
Wind.Speed	0.33460819
Maximum.Wind.Speed	0.07771593
Solar.Radiation	0.04144277
Heading	0.42056208
Battery.Life	0.89239275

PROGRAM 9:

```
library(FSelector)
cat("Step 1: Enter the data")
InputDataFileName="F:/beach.csv"
InputDataFileName
dataset<- read.csv(InputDataFileName) # Read the datafile
dataset<- dataset[sample(nrow(dataset)),] # Shuffle the data row wise.
cat("Step 2: Choose Target Variable")
target <- names(dataset)[1] # i.e. "Wet.Bulb.Temperature"
target
cat("Step 3: Create formula: symbolic creation of data -> ", target)
formula<- as.formula(paste(target, " ", paste(c(inputs), collapse =
"+")))
formula
e= rank.correlation(formula, data) # Correlation filter
e
```

9.2.1.4 Entropy-based Filter

This algorithm uses the correlation among the discrete attributes with continuous attributes to find out the weights of discrete attributes. There are three functions using this entropy-based filter, namely, information.gain, gain.ratio, and symmetrical.uncertainty. Program 10 is built with information gain, and the results are presented in Table 9.6. Program 11 uses the concept of gain ratio function, and the results are presented in Table 9.7. Program 12 illustrates the functionality of symmetrical.uncertainty, and the results are presented in Table 9.8. The syntax for these functions are as follows.

TABLE 9.5: Results of correlation filter

Feature	attr_importance
Humidity	0.08895382
Rain.Intensity	0.12602098
Interval.Rain	0.02935318
Total.Rain	0.86300361
Precipitation.Type	0.24828514
Wind.Direction	0.11143548
Wind.Speed	0.39384085
Maximum.Wind.Speed	0.02330741
Solar.Radiation	0.02706203
Heading	0.45769718
Battery.Life	0.82724846

information.gain(formula, data, unit)
gain.ratio(formula, data, unit)
symmetrical.uncertainty(formula, data, unit)
The arguments used in the function definition are described here.

1. formula: It is the description and working behind the algorithm.
2. data: It is the set of training data with the defined attributes for which the selection is to be made.
3. unit: It is the unit which is used for entropy computing. By default, it takes the value 'log'.

PROGRAM 10:

```
library(FSelector)
cat("Step 1: Enter the data")
InputDataFileName="F:/beach.csv"
InputDataFileName
dataset<- read.csv(InputDataFileName) # Read the datafile
dataset<- dataset[sample(nrow(dataset)),] # Shuffle the data row wise.
cat("Step 2: Choose Target Variable")
target <- names(dataset)[1] # i.e. "Wet.Bulb.Temperature"
target
cat("Step 3: Create formula: symbolic creation of data -> ", target)
formula<- as.formula(paste(target, " ", paste(c(inputs), collapse =
"+")))
formula
f= information.gain(formula, data) # Entropy-based filter
f
```

TABLE 9.6: Results of entropy-based filter

Feature	**attr_importance**
Humidity	0.10346771
Rain.Intensity	0.00000000
Interval.Rain	0.00000000
Total.Rain	0.76908219
Precipitation.Type	0.03076693
Wind.Direction	0.38151913
Wind.Speed	0.12602218
Maximum.Wind.Speed	0.06642477
Solar.Radiation	0.00000000
Heading	0.30532346
Battery.Life	0.61396799

PROGRAM 11:

```
library(FSelector)
cat("Step 1: Enter the data")
InputDataFileName="F:/beach.csv"
InputDataFileName
dataset<- read.csv(InputDataFileName) # Read the datafile
dataset<- dataset[sample(nrow(dataset)),] # Shuffle the data row
wise.
cat("Step 2: Choose Target Variable")
target <- names(dataset)[1] # i.e. "Wet.Bulb.Temperature"
target
cat("Step 3: Create formula: symbolic creation of data -> ", target)
formula<- as.formula(paste(target, " ", paste(c(inputs), collapse =
"+")))
formula
g= gain.ratio(formula, data) # Entropy-based filter
g
```

TABLE 9.7: Results of entropy-based filter

Feature	**attr_importance**
Humidity	0.1275170
Rain.Intensity	0.0000000
Interval.Rain	0.0000000
Total.Rain	0.5748086
Precipitation.Type	0.1505507
Wind.Direction	0.2113481
Wind.Speed	0.1932089
Maximum.Wind.Speed	0.1538750
Solar.Radiation	0.0000000
Heading	0.3716408
Battery.Life	0.6280430

PROGRAM 12:

```
library(FSelector)
cat("Step 1: Enter the data")
InputDataFileName="F:/beach.csv"
InputDataFileName
dataset<- read.csv(InputDataFileName) # Read the datafile
dataset<- dataset[sample(nrow(dataset)),] # Shuffle the data row
wise.
cat("Step 2: Choose Target Variable")
target <- names(dataset)[1] # i.e. "Wet.Bulb.Temperature"
target
cat("Step 3: Create formula: symbolic creation of data -> ", target)
formula<- as.formula(paste(target, " ", paste(c(inputs), collapse =
"+")))
formula
h= symmetrical.uncertainty(formula, data) # Entropy-based filter
h
```

TABLE 9.8: Results of entropy-based filter

Feature	attr_importance
Humidity	0.08548080
Rain.Intensity	0.00000000
Interval.Rain	0.00000000
Total.Rain	0.52186851
Precipitation.Type	0.03392537
Wind.Direction	0.22346297
Wind.Speed	0.11144039
Maximum.Wind.Speed	0.06508665
Solar.Radiation	0.00000000
Heading	0.25119237
Battery.Life	0.47465140

9.2.1.5 OneR Algorithm

This algorithm builds the association rule for each discrete attribute and uses that rule in the condition part to its weight. It makes use of OneR classifier. The program 13 is build for implementing this algorithm and the results are presented in Table 9.9. The syntax of this algorithm is as follows.
oneR(formula, data)
The arguments used in this algorithm are described here.

1. formula: It is the description of the working behind the algorithm.

2. data: It is the set of training data with the defined attributes for which the selection is made.

PROGRAM 13:

```
library(FSelector)
cat("Step 1: Enter the data")
InputDataFileName="F:/beach.csv"
InputDataFileName
dataset<- read.csv(InputDataFileName) # Read the datafile
dataset<- dataset[sample(nrow(dataset)),] # Shuffle the data row wise.
cat("Step 2: Choose Target Variable")
target <- names(dataset)[1] # i.e. "Wet.Bulb.Temperature"
target
cat("Step 3: Create formula: symbolic creation of data -> ", target)
formula<- as.formula(paste(target, " ", paste(c(inputs), collapse =
"+")))
formula
i= oneR(formula, data) # OneR algorithm
i
```

TABLE 9.9: Results of OneR algorithm

Feature	**attr_importance**
Humidity	0.540
Rain.Intensity	0.426
Interval.Rain	0.426
Total.Rain	0.800
Precipitation.Type	0.450
Wind.Direction	0.704
Wind.Speed	0.546
Maximum.Wind.Speed	0.480
Solar.Radiation	0.426
Heading	0.586
Battery.Life	0.706

9.2.1.6 RandomForest Filter

The algorithm behind the working of random.forest.importance method is the random forest. Program 14 is built using this method, and its results are presented in Table 9.10. The syntax of this function is as follows.
random.forest.importance(formula, data, importance.type)
The arguments used in the function definition are described here.

1. formula: It is the description of the working behind the algorithm.
2. data: It is the set of training data with the defined attributes for which the selection is made.
3. importance.type: Its value is either 1 or 2. It is the method used to measure the importance in which 1 means a decrease in accuracy and 2 means a decrease in node impurity.

PROGRAM 14:

```
library(FSelector)
cat("Step 1: Enter the data")
InputDataFileName="F:/beach.csv"
InputDataFileName
dataset<- read.csv(InputDataFileName) # Read the datafile
dataset<- dataset[sample(nrow(dataset)),] # Shuffle the data row
wise.
cat("Step 2: Choose Target Variable")
target <- names(dataset)[1] # i.e. "Wet.Bulb.Temperature"
target
cat("Step 3: Create formula: symbolic creation of data -> ", target)
formula<- as.formula(paste(target, " ", paste(c(inputs), collapse =
"+")))
formula
j= random.forest.importance(formula, data) # RandomForest filter
j
```

TABLE 9.10: Results of RandomForest filter

Feature	attr_importance
Humidity	24.950507
Rain.Intensity	2.721210
Interval.Rain	7.466201
Total.Rain	31.495858
Precipitation.Type	12.222609
Wind.Direction	32.901773
Wind.Speed	19.513976
Maximum.Wind.Speed	14.803433
Solar.Radiation	20.973147
Heading	32.984332
Battery.Life	47.901636

9.2.1.7 RReliefF Filter

The relief algorithm finds the weights of both continuous and discrete attributes by measuring the distance between instances. Program 15 uses this method, and its results are shown in Table 9.11. The syntax of this function is as follows: relief(formula, data, neighbours.count, sample.size)
The arguments used in the function definition are described here.

1. formula: It is the description and working of the algorithm.
2. data: It is the set of training data with the defined attributes for which the selection is made.
3. neighbours.count: It is the count of neighbors to be considered while computing each sampled instance.
4. sample.size: It is the size of instances in each sample.

PROGRAM 15:

```
library(FSelector)
cat("Step 1: Enter the data")
InputDataFileName="F:/beach.csv"
InputDataFileName
dataset<- read.csv(InputDataFileName) # Read the datafile
dataset<- dataset[sample(nrow(dataset)),] # Shuffle the data row
wise.
cat("Step 2: Choose Target Variable")
target <- names(dataset)[1] # i.e. "Wet.Bulb.Temperature"
target
cat("Step 3: Create formula: symbolic creation of data -> ", target)
formula<- as.formula(paste(target, " ", paste(c(inputs), collapse =
"+")))
formula
k= relief(formula, data) #RReliefF filter
k
```

TABLE 9.11: Results of RReliefF filter

Feature	**attr_importance**
Humidity	0.037686511
Rain.Intensity	0.000000000
Interval.Rain	0.053114263
Total.Rain	0.120525966
Precipitation.Type	0.000000000
Wind.Direction	0.363366182
Wind.Speed	0.045568396
Maximum.Wind.Speed	0.077139625
Solar.Radiation	0.161974706
Heading	0.070068755
Battery.Life	0.004803814

9.2.2 Recursive Feature Elimination

After computing the importance of each feature, the wrapper method *recursive feature elimination* is used. This method is based on the performance of the model. The model is trained with the selected features and the performance is evaluated. This step is repeated until the desired performance is achieved and the features selected for the optimal performance is returned. The code for RFE is depicted in Program 6. The result of the program is presented in Figure 9.5. The working of this method is elaborated below:

1. Train the model with all the predictors.
2. Calculate the performance of the model.
3. Calculate the importance of each feature.
4. Repeat these steps:
   - $\forall$ For the subset S_1.... S_i.
   - Keep the most important features.
   - Discard the irrelevant features.
   - Train the model with training set over S_i.
   - Evaluate the model performance.
   - Again calculate the importance of each predictor.
5. Calculate the model performance with the best subset S_i.
6. List the features in the best subset.

Program 6:

```
set.seed(7)
library(mlbench)
library(caret)
data<- read.csv("F:/aaisha's paper/book/Format/experiments/
beach1.csv")
control<- rfeControl(functions=rfFuncs, method="cv", number=10)
# running the RFE algorithm
results<- rfe(data[,1:12], data[,12], sizes=c(1:12), rfeControl=control)
# summarizing the results
print(results)
predictors(results)
plot(results, type=c("g", "o"))
```

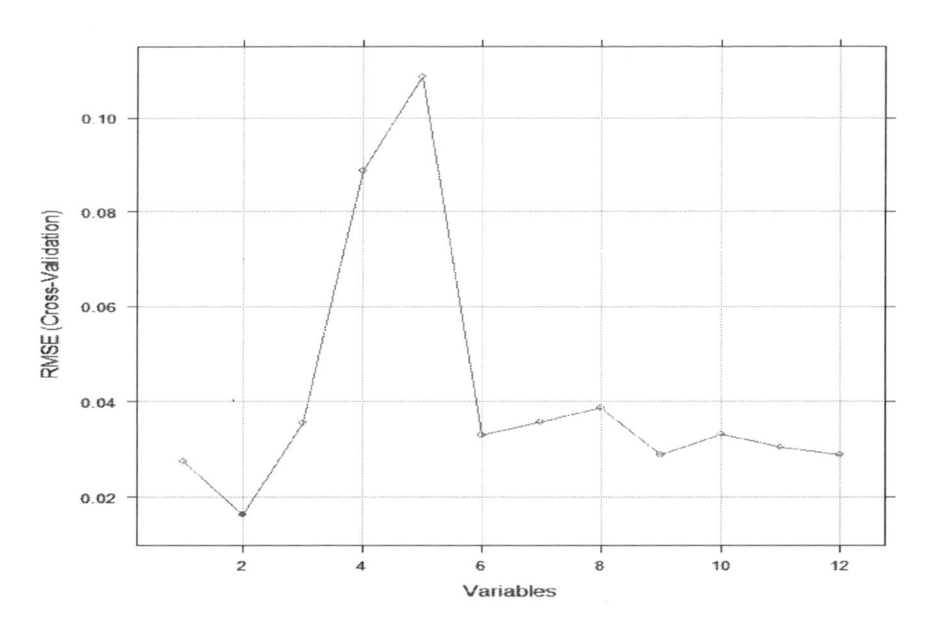

FIGURE 9.5: Wrapper method for feature elimination

OUTPUT:

Recursive feature selection
Outer resampling method: Cross-Validated (10 fold)
Resampling performance over subset size:
Variables RMSE Rsquared MAE RMSESD RsquaredSD MAESD Selected
1 0.02757 0.9994 0.02062 0.027035 0.00064974 0.0214619
2 0.01628 0.9999 0.01072 0.001224 0.00001772 0.0006904
3 0.03547 0.9993 0.01786 0.018762 0.00069679 0.0040235
4 0.08862 0.9977 0.05944 0.025599 0.00170378 0.0156416
5 0.10875 0.9971 0.07883 0.026934 0.00193064 0.0236303
6 0.03296 0.9994 0.01588 0.019406 0.00065571 0.0046859
7 0.03568 0.9993 0.01743 0.021286 0.00079294 0.0049197
8 0.03874 0.9992 0.01897 0.023467 0.00096142 0.0050884
9 0.02896 0.9995 0.01186 0.016822 0.00050179 0.0035232
10 0.03310 0.9994 0.01334 0.018844 0.00062634 0.0039566
11 0.03053 0.9995 0.01363 0.016734 0.00052578 0.0041760
12 0.02881 0.9995 0.01093 0.018857 0.00058098 0.0038946
The top 2 variables (out of 2):
Wet.Bulb.Temperature, Battery.Life
>predictors(results)
[1] "Wet.Bulb.Temperature" "Battery.Life"
>plot(results, type=c("g", "o"))

9.3 Machine Learning Models

To have the machine learn and predict the unseen pattern requires a procedure to be followed. We program that procedure and call it a machine learning model. There are more than 200 machine learning models developed to date. Discussing all the models is behind the scope of this book. We discuss a few of these which are implemented with the previously mentioned IoT dataset as presented in Table 9.12. We have implemented the multiple models and the results are presented in Tables 9.13, 9.14, 9.15, 9.16, 9.17, and 9.18. After learning the various steps of data pre processing, feature extraction, and feature selection, let's learn the working of the machine learning model. So, now the program written below does not include the initial steps, assuming the data is already pre processed, followed by feature extraction and feature selection. Implementing machine learning models requires the following steps:

1. Include all the libraries, such as rpart, FSelector required for the execution of the functions being called in the consecutive steps of the program.
2. Declare the variables required to handle the functionality of the model. Three variables are declared, *modelName* to store the name of model, *InputData* to store the path of the file to be taken as a data file and the *training* to store the training ratio.
3. Load the data from the source file as mentioned in the input data.
4. Count the rows in the dataset to ensure the successful loading of complete data.
5. Select the target attribute for which the prediction is to be made.
6. Mark the attributes as input attribute other than the target attribute.
7. Select the training data from the existing data.
8. Select the testing data after selecting the training data.
9. Design the working of a model and define the function of the model.
10. Perform the prediction using the testing data.
11. Extract the actual instances from the dataset.
12. Evaluate the performance of the model, by computing the evaluation parameters, i.e., correlation, RSquare, RMSE, accuracy.
13. Write the results in the file.

```
cat("Step 1: Library Inclusion") %Includes the libraries required by the
                                 % functions used in the program
library(rpart)
library(FSelector)
cat("Step 2: Variable Declaration")
modelName<- "decisionTree"        %Declare the variables
InputData="F:/beach1.csv"         %The path of .csv file in your system
training = 50                     %This defines the ratio of training data
                                  %and ratio of testing data is
                                  %100 - training
cat("Step 3: Data Loading")       %This step loads the file into the
                                  %R workspace
dataset<- read.csv(InputDataFileName)     %read.csv is the function used
                                          %to read the csv file
dataset<- dataset[sample(nrow(dataset)),] %Shuffle the data row wise

cat("Step 4: Counting dataset")        %count the number of rows in data
totalDataset<- nrow(dataset)

cat("Step 5: Choose Target Attribute")  %Set the target attribute among
                                        %the attributes in data
target <- names(dataset)[1]

cat("Step 6: Choose Input Attributes")  %Set the other attributes except
                                        %the target attribute as the
                                        %tinputs
inputs<- setdiff(names(dataset),target)

cat("Step 7: Select training dataset")  %select the training data from
                                        %the input data
trainDataset<- dataset[1:(totalDataset * training/100),c(inputs, target)]

cat("Step 8: Select testing dataset")   %select the testing data from the
                                        %input data
testDataset<- dataset[(totalDataset * training/100):totalDataset,c(inputs,
target)]

cat("Step 9: Model Building -> ", modelName)
formula<- as.formula(paste(target, " ", paste(c(inputs), collapse = "+")))
                                    %methodology of the working of model

model<- rpart(formula, trainDataset, parms=list(split="information"),
control=rpart.control(usesurrogate=0, maxsurrogate=0)) %model definition
cat("Step 10: Prediction using -> ", modelName)     %Predicting
Predicted <- predict(model, testDataset)
```

TABLE 9.13: Performance of H with respect to models and samples

Set	M1	M2	M3	M4	M5	M6	M7	M8	M9	M10	M11	M12	M13	M14	M15
S1	0.969	0.977	1	0.852	0.798	0.766	0.99	0.835	0.897	0.97	0.862	0.951	0.972	0.914	0.986
S2	0.985	0.96	0.986	0.897	0.683	0.856	0.97	0.847	0.873	0.985	0.918	0.959	0.959	0.872	0.945
S3	1	0.943	0.965	0.798	0.775	0.867	0.968	0.835	0.795	0.942	0.955	0.963	0.935	0.865	0.945
S4	0.961	0.926	0.917	0.937	0.826	0.853	0.971	0.89	0.762	0.865	0.904	0.958	0.958	0.945	0.959
S5	0.878	0.959	0.927	0.819	0.773	0.886	0.98	0.829	0.695	0.932	0.877	0.927	0.899	0.881	0.887
S6	0.907	0.978	0.892	0.767	0.701	0.873	0.979	0.779	0.732	0.875	0.737	0.928	0.878	0.815	0.904
S7	0.979	1	0.981	0.861	0.659	0.848	0.969	0.902	0.853	0.899	0.768	0.978	0.901	0.777	0.906
S8	1	0.346	0.986	0.223	0.104	0	0.959	0.33	0.76	0.113	0.714	0.986	0.351	0.07	1
S9	1	0.988	1	0.529	0.438	0.593	0.98	0.657	0.772	0.653	0.879	0.988	0.664	0.661	0.986
S10	0.989	0.98	0.986	0.77	0.699	0.801	0.99	0.956	0.807	0.845	0.847	0.934	0.882	0.741	1
S11	0.928	0.894	0.925	0.874	0.521	0.861	0.961	0.817	0.778	0.971	0.905	0.949	0.935	0.945	0.942
Avg.	0.963	0.904	0.960	0.757	0.634	0.745	0.974	0.7888	0.793	0.822	0.851	0.956	0.848	0.771	0.950

Note: M1= Bagged MARS, M2= Bagged CART, M3= Bagged Model, M4= Bayesian Generalized Linear Model, M5= Boosted Linear Model, M6= Conditional Inference Tree, M7= eXtreme Gradient Boosting, M8= Generalized Linear Model with Stepwise Feature Selection, M9= k-Nearest Neighbors, M10= Least Angle Regression, M11= Parallel Random Forest, M12= partDSA, M13= Partial Least Squares, M14= Partial Least Squares Generalized Linear Models, M15= Tree-Based Ensembles, S1-S12= Sample sets

TABLE 9.14: Performance of Gini with respect to models and samples

Set	M1	M2	M3	M4	M5	M6	M7	M8	M9	M10	M11	M12	M13	M14	M15
S1	0.987	0.999	1	0.966	0.899	0.859	0.993	0.927	0.981	0.974	0.935	0.97	0.986	0.984	0.991
S2	0.98	0.988	0.982	0.963	0.789	0.913	0.979	0.921	0.95	0.979	0.943	0.973	0.948	0.96	0.963
S3	1	0.965	0.38	0.932	0.841	0.919	0.978	0.921	0.907	0.929	0.962	0.977	0.935	0.926	0.962
S4	0.995	0.947	0.919	0.979	0.889	0.909	0.981	0.898	0.874	0.895	0.935	0.971	0.945	0.947	0.972
S5	0.925	0.976	0.977	0.893	0.838	0.924	0.989	0.902	0.872	0.908	0.94	0.965	0.858	0.901	0.921
S6	0.92	0.999	0.967	0.856	0.741	0.915	0.985	0.889	0.861	0.925	0.859	0.995	0.834	0.909	0.94
S7	0.999	1	0.999	0.914	0.776	0.899	0.978	0.938	0.93	0.857	0.829	0.988	0.863	0.825	0.934
S8	1	0.367	0.997	0.095	0.006	0	0.972	0.097	0.892	0.065	0.842	0.99	0.075	0.108	1
S9	1	0.992	1	0.45	0.483	0.696	0.986	0.562	0.874	0.495	0.96	0.998	0.52	0.61	0.99
S10	1	0.999	0.991	0.759	0.738	0.859	0.993	0.937	0.937	0.842	0.957	0.945	0.847	0.824	1
S11	0.908	0.918	0.901	0.886	0.676	0.903	0.976	0.876	0.899	0.959	0.958	0.97	0.942	0.951	0.96
Avg.	0.974	0.922	0.973	0.790	0.697	0.799	0.982	0.806	0.907	0.802	0.92	0.976	0.795	0.813	0.966

Note: M1= Bagged MARS, M2= Bagged CART, M3= Bagged Model, M4= Bayesian Generalized Linear Model, M5= Boosted Linear Model, M6= Conditional Inference Tree, M7= eXtreme Gradient Boosting, M8= Generalized Linear Model with Stepwise Feature Selection, M9= k-Nearest Neighbors, M10= Least Angle Regression, M11= Parallel Random Forest, M12= partDSA, M13= Partial Least Squares, M14= Partial Least Squares Generalized Linear Models, M15= Tree-Based Ensembles, S1-S12= Sample sets

TABLE 9.15: Performance of AUC with respect to models and samples

Set	M1	M2	M3	M4	M5	M6	M7	M8	M9	M10	M11	M12	M13	M14	M15
S1	0.994	1	1	0.983	0.949	0.929	0.997	0.963	0.99	0.987	0.967	0.985	0.993	0.992	0.995
S2	0.99	0.994	0.991	0.981	0.895	0.956	0.99	0.96	0.975	0.99	0.972	0.986	0.974	0.98	0.981
S3	1	0.982	0.99	0.966	0.921	0.959	0.989	0.96	0.954	0.964	0.981	0.988	0.968	0.963	0.981
S4	0.997	0.973	0.96	0.99	0.944	0.954	0.99	0.949	0.937	0.947	0.968	0.985	0.972	0.974	0.986
S5	0.963	0.988	0.989	0.946	0.919	0.962	0.994	0.951	0.936	0.954	0.97	0.982	0.929	0.951	0.96
S6	0.96	0.999	0.984	0.928	0.87	0.957	0.993	0.945	0.93	0.962	0.929	0.997	0.917	0.955	0.97
S7	1	1	1	0.957	0.888	0.95	0.989	0.969	0.965	0.929	0.914	0.994	0.931	0.913	0.967
S8	1	0.684	0.999	0.547	0.503	0.5	0.986	0.548	0.946	0.532	0.921	0.995	0.538	0.554	1
S9	1	0.996	1	0.725	0.741	0.848	0.993	0.781	0.937	0.748	0.98	0.999	0.76	0.805	0.995
S10	1	1	0.996	0.88	0.869	0.929	0.997	0.968	0.969	0.921	0.978	0.973	0.924	0.912	1
S11	0.954	0.959	0.951	0.943	0.838	0.952	0.988	0.938	0.949	0.98	0.979	0.985	0.971	0.976	0.98
Avg.	0.987	0.961	0.987	0.895	0.848	0.899	0.991	0.902	0.953	0.901	0.959	0.988	0.897	0.906	0.983

Note: M1= Bagged MARS, M2= Bagged CART, M3= Bagged Model, M4= Bayesian Generalized Linear Model, M5= Boosted Linear Model, M6= Conditional Inference Tree, M7= eXtreme Gradient Boosting, M8= Generalized Linear Model with Stepwise Feature Selection, M9= k-Nearest Neighbors, M10= Least Angle Regression, M11= Parallel Random Forest, M12= partDSA, M13= Partial Least Squares, M14= Partial Least Squares Generalized Linear Models, M15= Tree-Based Ensembles, S1-S12= Sample sets

TABLE 9.16: Performance of AUCH with respect to models and samples

Set	M1	M2	M3	M4	M5	M6	M7	M8	M9	M10	M11	M12	M13	M14	M15
S1	0.997	1	1	0.986	0.97	0.929	0.997	0.973	0.992	0.993	0.972	0.985	0.996	0.994	0.995
S2	0.995	0.994	0.995	0.986	0.921	0.956	0.99	0.969	0.978	0.995	0.972	0.986	0.986	0.985	0.981
S3	1	0.984	0.995	0.974	0.947	0.959	0.989	0.969	0.96	0.98	0.987	0.988	0.982	0.975	0.981
S4	0.998	0.973	0.978	0.994	0.969	0.954	0.99	0.967	0.939	0.964	0.972	0.985	0.986	0.987	0.986
S5	0.972	0.988	0.992	0.96	0.944	0.962	0.996	0.964	0.941	0.977	0.977	0.982	0.965	0.974	0.96
S6	0.978	0.999	0.988	0.944	0.912	0.957	0.993	0.961	0.934	0.974	0.934	0.997	0.958	0.965	0.975
S7	1	1	1	0.969	0.917	0.95	0.989	0.98	0.967	0.964	0.918	0.994	0.966	0.945	0.976
S8	1	0.736	0.999	0.696	0.639	0.5	0.986	0.749	0.948	0.659	0.93	0.995	0.747	0.604	1
S9	1	0.996	1	0.839	0.828	0.848	0.993	0.877	0.937	0.874	0.983	0.999	0.879	0.887	0.995
S10	1	1	0.998	0.935	0.926	0.929	0.997	0.984	0.973	0.954	0.985	0.978	0.962	0.938	1
S11	0.974	0.963	0.973	0.957	0.866	0.952	0.992	0.955	0.951	0.99	0.982	0.985	0.983	0.983	0.98
Avg.	0.992	0.966	0.992	0.930	0.894	0.899	0.992	0.940	0.956	0.938	0.964	0.988	0.946	0.930	0.984

Note: M1= Bagged MARS, M2= Bagged CART, M3= Bagged Model, M4= Bayesian Generalized Linear Model, M5= Boosted Linear Model, M6= Conditional Inference Tree, M7= eXtreme Gradient Boosting, M8= Generalized Linear Model with Stepwise Feature Selection, M9= k-Nearest Neighbors, M10= Least Angle Regression, M11= Parallel Random Forest, M12= partDSA, M13= Partial Least Squares, M14= Partial Least Squares Generalized Linear Models, M15= Tree-Based Ensembles, S1-S12= Sample sets

TABLE 9.17: Performance of Youden with respect to models and samples

Set	M1	M2	M3	M4	M5	M6	M7	M8	M9	M10	M11	M12	M13	M14	M15
S1	0.922	0.95	0.88	0.745	0.79	0.859	0.993	0.763	0.92	0.457	0.899	0.97	0.771	0.69	0.991
S2	0.918	0.971	0.922	0.755	0.673	0.913	0.979	0.721	0.896	0.546	0.943	0.973	0.731	0.699	0.963
S3	0.899	0.96	0.953	0.721	0.743	0.919	0.978	0.748	0.831	0.495	0.964	0.977	0.649	0.739	0.962
S4	0.924	0.947	0.931	0.835	0.8	0.909	0.981	0.691	0.775	0.748	0.934	0.971	0.75	0.738	0.972
S5	0.884	0.973	0.897	0.699	0.777	0.924	0.986	0.667	0.755	0.56	0.894	0.953	0.663	0.714	0.921
S6	0.926	0.888	0.907	0.703	0.759	0.915	0.985	0.697	0.778	0.657	0.809	0.934	0.713	0.685	0.933
S7	0.943	1	0.984	0.75	0.734	0.899	0.978	0.686	0.881	0.429	0.836	0.987	0.735	0.702	0.936
S8	0.919	0.462	0.943	0.183	0.013	0	0.972	0.198	0.719	0	0.789	0.99	0.21	0	1
S9	0.98	0.992	0.97	0.549	0.223	0.696	0.986	0.571	0.838	0.534	0.885	0.993	0.54	0.676	0.99
S10	0.96	0.977	0.971	0.648	0.255	0.859	0.993	0.768	0.8	0.714	0.872	0.957	0.79	0.614	1
S11	0.918	0.926	0.914	0.716	0.39	0.903	0.899	0.735	0.788	0.571	0.912	0.964	0.703	0.701	0.96
Avg.	0.926	0.913	0.933	0.664	0.557	0.799	0.975	0.622	0.816	0.519	0.885	0.969	0.621	0.632	0.966

Note: M1= Bagged MARS, M2= Bagged CART, M3= Bagged Model, M4= Bayesian Generalized Linear Model, M5= Boosted Linear Model, M6= Conditional Inference Tree, M7= eXtreme Gradient Boosting, M8= Generalized Linear Model with Stepwise Feature Selection, M9= k-Nearest Neighbors, M10= Least Angle Regression, M11= Parallel Random Forest, M12= partDSA, M13= Partial Least Squares, M14= Partial Least Squares Generalized Linear Models, M15= Tree-Based Ensembles, S1-S12= Sample sets

TABLE 9.18: Performance of accuracy with respect to models and samples

Set	M1	M2	M3	M4	M5	M6	M7	M8	M9	M10	M11	M12	M13	M14	M15
S1	96.79	98	94.4	89.2	91.6	91.79	99.71	89.2	96.4	79.6	95.79	98.4	91.2	87.6	99.6
S2	96.8	98.8	96.8	88.8	85.2	95.16	99.14	88.4	95.6	82.4	97.68	98.8	88.4	86.4	98.4
S3	96	98.4	98	88.8	88	95.58	99.14	88.4	92.8	80	98.53	98.8	84.4	90.8	98.4
S4	96.81	98.01	97.21	90.8	92.03	95.17	99.15	88.45	90.44	89.64	97.27	98.8	89.64	88.84	98.8
S5	95.6	98.8	96	86.8	90	96.63	99.43	85.6	89.6	80.8	95.16	97.6	86.8	87.2	96.8
S6	96.81	95.6	96	87.2	89.6	96.21	99.43	87.2	89.6	85.6	92	96.4	84.8	88	97.2
S7	97.6	100	99.2	88.8	88	95.37	99.14	86.8	94.8	77.6	93.26	99.2	89.2	87.2	97.2
S8	96.79	77.8	97.59	67.47	60.24	59.2	98.85	65.86	87.95	57.43	91.33	99.6	66.67	59.04	100
S9	99.2	99.67	98.8	82.4	68	87.79	99.43	82	92.8	80.8	94.74	99.6	81.6	86.4	99.6
S10	98.41	99	98.01	84.46	70.92	94.33	99.72	91.24	90.84	87.25	94.54	98.01	91.24	83.67	100
S11	96.8	97	96.8	88.4	74.4	96	94.57	88.4	90.8	83.2	96.42	98.4	88	87.2	98.4
Avg.	97.05	96.46	97.16	85.73	81.63	91.2	98.88	85.59	91.96	80.40	95.15	98.51	85.63	84.76	98.58

Note: M1= Bagged MARS, M2= Bagged CART, M3= Bagged Model, M4= Bayesian Generalized Linear Model, M5= Boosted Linear Model, M6= Conditional Inference Tree, M7= eXtreme Gradient Boosting, M8= Generalized Linear Model with Stepwise Feature Selection, M9= k-Nearest Neighbors, M10= Least Angle Regression, M11= Parallel Random Forest, M12= partDSA, M13= Partial Least Squares, M14= Partial Least Squares Generalized Linear Models, M15= Tree-Based Ensembles, S1-S12= Sample sets

TABLE 9.19: Summary of performance of all the models

M	1	2	3	4	5	6	7	8	9	10	11	12	13	14	15	16	17	18	19	20
M1	0.963	0.974	0.987	0.992	0.973	0.011	0.012	0.889	0.980	0.029	0.927	0.998	0.998	0.927	0.927	0.001	0.961	0.926	97.055	132.6
M2	0.904	0.922	0.961	0.966	0.926	0.030	0.035	0.764	0.939	0.035	0.913	1	1	0.913	0.913	0	0.946	0.913	96.46	7.182
M3	0.960	0.973	0.987	0.992	0.970	0.0127	0.014	0.910	0.978	0.027	0.934	0.999	0.999	0.934	0.934	0	0.965	0.933	97.16	58.20
M4	0.757	0.790	0.895	0.930	0.821	0.078	0.085	0.498	0.846	0.137	0.676	0.987	0.960	0.676	0.676	0.012	0.786	0.664	85.73	3.21
M5	0.634	0.697	0.848	0.894	0.728	0.120	0.135	0.275	0.719	0.181	0.562	0.995	0.9	0.562	0.562	0.004	0.668	0.557	81.63	12.31
M6	0.745	0.799	0.899	0.899	0.799	0.087	0.096	0.699	0.742	0.087	0.834	0.965	0.949	0.834	0.834	0.034	0.928	0.799	91.20	3.437
M7	0.974	0.982	0.991	0.992	0.982	0.007	0.008	1	0.995	0.011	0.982	0.993	0.991	0.982	0.982	0.006	0.986	0.975	**98.88**	66.07
M8	0.788	0.806	0.902	0.940	0.845	0.068	0.074	0.555	0.873	0.170	0.716	0.906	0.937	0.716	0.716	0.094	0.796	0.622	85.595	5.99
M9	0.793	0.907	0.953	0.956	0.844	0.068	0.074	0.472	0.870	0.080	0.846	0.970	0.951	0.846	0.846	0.03	0.894	0.816	91.96	3.649
M10	0.822	0.802	0.901	0.938	0.867	0.060	0.0643	0.578	0.858	0.209	0.610	0.908	0.946	0.610	0.610	0.091	0.710	0.519	80.39	2.243
M11	0.8514	0.92	0.959	0.964	0.891	0.044	0.052	0.654	0.909	0.048	0.891	0.993	0.990	0.891	0.891	0.006	0.937	0.885	95.156	3.902
M12	0.956	0.976	0.988	0.988	0.969	0.014	0.014	0.992	0.994	0.014	0.985	0.984	0.98	0.985	0.985	0.015	0.982	0.969	**98.51**	4.56
M13	0.848	0.795	0.897	0.946	0.887	0.047	0.054	0.455	0.893	0.171	0.712	0.909	0.942	0.712	0.712	0.090	0.795	0.621	85.63	2.328
M14	0.771	0.813	0.906	0.930	0.825	0.077	0.084	0.525	0.828	0.151	0.636	0.995	0.99	0.636	0.636	0.004	0.820	0.632	84.75	33.05
M15	0.950	0.966	0.983	0.984	0.966	0.014	0.016	0.860	0.987	0.014	0.966	1	1	0.966	0.966	0	0.982	0.966	**98.58**	139.05

TABLE 9.20: Performance of ensemble model

M	1	2	3	4	5	6	7	8	9	10	11	12	13	14	15	16	17	18	19	20
EM	0.985	0.99	0.995	0.995	0.99	0.004	0.005	1	0.999	0.004	1	0.99	1	1	0.99	0	0.995	0.99	99.3	746.9

TABLE 9.21: Ten rounds of cross validation

R	1	2	3	4	5	6	7	8	9	10	11	12	13	14	15	16	17	18	19	20
1	0.985	0.99	0.995	0.995	0.99	0.004	0.005	1	0.999	0.004	0.99	1	1	0.99	0.99	0	0.995	0.99	99.6	754.1
2	0.986	0.991	0.995	0.995	0.991	0.004	0.005	1	1	0.004	0.991	1	1	0.991	0.991	0	0.995	0.991	99.6	681.6
3	0.986	0.991	0.995	0.995	0.991	0.004	0.005	1	1	0.004	0.991	1	1	0.991	0.991	0	0.995	0.991	99.6	729.3
4	0.97	0.979	0.989	0.989	0.979	0.008	0.01	1	0.999	0.008	0.979	1	1	0.979	0.979	0	0.989	0.979	99.2	18015
5	0.971	0.98	0.99	0.99	0.98	0.008	0.01	1	0.999	0.008	0.98	1	1	0.98	0.98	0	0.99	0.98	99.2	826
6	0.96	0.97	0.985	0.989	0.97	0.008	0.01	1	1	0.008	0.97	1	1	0.97	0.96	0	0.99	0.97	99.14	576
7	0.969	0.978	0.989	0.989	0.978	0.009	0.011	1	0.999	0.009	0.978	1	1	0.978	0.978	0	0.989	0.978	99.14	60.49
8	0.98	0.989	0.994	0.996	0.986	0.006	0.007	1	0.986	0.006	0.986	1	1	0.986	0.986	0	0.993	0.986	99.2	70.9
9	0.986	0.99	0.995	0.995	0.99	0.004	0.005	1	1	0.004	0.99	1	1	0.99	0.99	0	0.995	0.99	99.6	37.5
10	0.988	0.998	0.999	0.999	0.993	0.004	0.003	0.999	1	0.004	1	0.993	0.991	1	1	0.007	0.995	0.993	99.6	34.91

Note*: 1= H measure, 2=Gini, 3=AUC, 4=AUCH, 5=KS, 6=MER, 7=MWL, 8=Spec.Sens95, 9=Sens.Spec95, 10=ER , 13=Precision, 14= Recall, 15=TPR, 16=FPR, 17= F-measure, 18=Youden, 19=Accuracy, 20=Execution time

Note**: M1= Bagged MARS, M2= Bagged CART, M3= Bagged Model, M4= Bayesian Generalized Linear Model, M5= Boosted Linear Model, M6= Conditional Inference Tree, M7= eXtreme Gradient Boosting, M8= Generalized Linear Model with Stepwise Feature Selection, M9= k-Nearest Neighbors, M10= Least Angle Regression, M11= Parallel Random Forest, M12= partDSA, M13= Partial Least Squares, M14= Partial Least Squares Generalized Linear Models, M15= Tree-Based Ensembles

the branches. Would it be the correct approach if the sets formed are small in the size? Most probably no, because if the model is trained if less number of instances than the learning power of the model decreases. If the learning is not good, then the model would not able to predict accurately with the testing set, which in return would lead to less accuracy. There is no such technique which would not return with an error. The error would surely occur, but the objective of each model is to diminish the error and improve the accuracy.

Now the question is, how we can expect the error rate to variate in same model but with different data. Suppose we evaluate the error rate by averaging the error estimated by all the classifiers. We illustrate this with an example. Suppose, you prefer computer languages more than other subjects of the computer. At the time of practice, we are giving the example of programming in other subjects like Database Management System. It means you are biased towards programming. Such an error can only be estimated. The second kind of error occurs when the model is not trained with the all possible patterns/combinations of instances. During testing, when such a new instance occurs, it will generate an error because the system is unaware of such a situation. This is called a variance of the classifier. So, we can conclude the biasness and variance; both together form the total error. The approach of combining actually decreases the variance. Bagging resamples the original data. It shuffles the data in such a manner that few instances get replaced, and few get deleted. The original data and the data produced during bagging do not differ much but are not exactly the same.

Bagging can also be applied to learning methods for numeric prediction, for example, model trees. The only difference is that, instead of voting on the outcome, the individual predictions, being real numbers, are averaged. The bias-variance decomposition can be applied to numeric prediction as well by decomposing the expected value of the mean-squared error of the predictions on fresh data. Bias is defined as the mean-squared error expected when averaging over models built from all possible training datasets of the same size, and variance is the component of the expected error of a single model that is due to the particular training data it was built from. It can be shown theoretically that averaging over multiple models built from independent training sets always reduces the expected value of the mean-squared error.

9.4.2 Boosting

Combining the models is significant when they differ in their working and rectify the data correctly. Combining the models is exactly the same as seeking advice from an expert. Because every model is expert in their own domain, combining does not mean copying the result' it means utilizing each other's specialization and experience. Boosting also uses the concepts of voting and averaging to combine the different models. Like bagging, it is essential to combine the similar kind of classifiers. The difference between bagging and boosting is that in bagging individual models are executed differently, and thus, the results are combined. However, in boosting, each new model is influenced by the performance of those built previously. In boosting, a model's contribution is influenced by its performance rather than weight.

9.5 Ensemble Approach

An ensemble is a technique which is adopted to improve the performance of classifiers. In the proposed framework, five machine learning models are an ensemble. The ensemble approach is built in such a manner that it improves the performance of each classifier. The following steps are used for the building of a new ensemble approach.

1. Collect the data to be fed into the classifier.
2. Distribute the data into the sets n, where n is the number of classifiers.
3. Train each classifier with one of the dataset.
4. Produce the hypothesis(h) from each classifier.
5. Combine the hypothesis.
6. Evaluate performance.

Many ensemble methods exist in literature. A few of them are listed below:

- Unweighted voting: Each classifier not only produces the classification decision but the class probability estimate. the estimates produced by all the classifiers are combined by the Equation 9.1. In this equation, h_l is the classifier which results in true prediction for k at a data point x.

$$P(f(x) = k) = \frac{1}{L}\sum_{l=1}^{L} P(f(x) = k|h_l) \tag{9.1}$$

- Least squares: This method is suitable for regression problems. This method works to target the maximum weights which improve the accuracy of the ensemble model. The principle applied states that the variance of estimate by h_l is inversely proportional to the weight of h_l [26].
- Likelihood combination: This method is suitable for the classification problem. In this, the accuracy of each classifier is computed to calculate the independent classifier weight. The methodology of this method uses the prior distribution P(h_l), which is multiplied with estimated likelihood P($S|h_l$) [2].

- Gating networks: It is that method of combining classifiers which accept the input x and produces the output w_l. Equation 9.2 produces the dot product of input x with parameter V. After computing Z, the weights W are calculated as in Equation 9.3. The output of Eq. 9.3 W_l is known as the *soft-max* of output of Equation Z_l [14].

$$Z_l = V_{l^T} * x \tag{9.2}$$

$$W_l = \frac{e^{Z_l}}{e^{\sum_u Z_u}} \tag{9.3}$$

- Stacking: It is implemented with leave-one-out cross-validation. For each classifier, and for each training set, a combinational hypothesis is produced. In the next iteration, the same procedure is followed except the last one [36].

- Divide the data and combine the hypothesis (DDCH): It reduces the load of the ranking module of the search engine. The experiments are conducted on the webspam dataset. In the Equation 9.4, n refers to number of classifiers; h_i is the hypothesis generated by each classifier.

$$S = \sum_{i=1}^{n} \frac{P|h_i}{n} \tag{9.4}$$

9.6 Summary and What's Next

This chapter explained the crucial step of feature engineering in detail. The feature reduction and feature selection techniques were covered along with the R code. It listed all the filter methods and wrapper methods along with the arguments. Writing and coding the machine learning model were also explained. The decision tree model has been built. Improving the performance of the model is described with the underlying techniques of boosting, bagging, and ensemble approaches.

In the next chapter, we will learn how to evaluate the performance of machine learning models.

9.7 Exercises

Questions

1. What is meant by feature reduction? What are the various techniques of feature reduction?
2. What is meant by feature selection? What are the different methods of feature selection?
3. What are the different ways to compute PCA?
4. What are the various steps for implementing the machine learning model?
5. How do we define the working of the model.
6. How do we evaluate the model? What are the various parameters calculated?
7. What is meant by bagging?
8. What is meant by boosting?
9. What are the different ensemble approaches?
10. What is meant by stacking?

Multiple Choice Questions

1: Which technique is known as the data compression technique?

(a) PCA **(b)** cfs

(c) rfe **(d)** None of the above

2: Which method is used to compute PCA ?

(a) prcomp() **(b)** primcomp()

(c) pca() **(d)** All of the above

3: Which method is used to evaluate the importance of features?

(a) Chi-squared filter

(b) Consistency method

(c) Correlation filter

(d) All of above

4: When we define the training and testing ratio?

(a) At the time of data loading

(b) Before data loading

(c) While selecting the dataset

(d) While defining the model

5: How do we evaluate the model?

(a) By testing it against testing dataset

(b) By computing its computational cost

(c) By comparing actual and predicted

(d) All of them

6: How do you we improve the performance of a model?

(a) Bagging approach

(b) Boosting approach

(c) Ensemble approach

(d) All of the above

7: Which package needs to be installed for implementing the bagged model?

(a) Earth

(b) Arm

(c) Caret

(d) Lars

8: Which package should be installed for implementing the partial least squares model?

(a) Pls **(b)** PlsRglm

(c) Mars **(d)** Party

9: What is the inbuilt function in R for computing the correlation?

(a) Correlation **(b)** cor

(c) Spearman **(d)** None of above

10: What are the tuning parameters of the KNN model?

(a) Kmax **(b)** Distance

(c) Kernel **(d)** All of the above

11: What is the output of the recursive feature elimination round?

(a) A subset consisting of less than n features **(b)** Returns the importance score of each feature

(c) The reduced dataset **(d)** Best possible features forming a subset

---

12: Which function of R is used for counting the rows in the data?

(a) read.csv **(b)** write.csv

(c) nrow **(d)** cat

13: Which command of R is used for defining the formula of a model?

(a) as.formula

(b) formula

(c) model

(d) as.model

14: Which method is used in tree-based ensembles?

(a) Ctree

(b) nodeHarvest

(c) Treebag

(d) None of the above

15: Which ensemble method was developed to reduce the ranking module of a search engine?

(a) Stacking

(b) Divide the data and combine the hypothesis (DDCH)

(c) Likelihood combination

(d) Gating networks

16: The process of removing redundant features is known as:

(a) Feature selection

(b) Feature reduction

(c) Principal component analysis

(d) None of the above

17: Which library is required to implement the filter methods of feature selection?

(a) FSelector

(b) rpart

(c) ggplot

(d) None of the above

18: Which filter method works on the principle of best first search?

(a) OneR filter

(b) Consistency-based filter

(c) Entropy-based filter

(d) None of the above

19: Recursive feature elimination is a filter method which accepts the arguments in the form of:

(a) List

(b) Vector

(c) Matrix

(d) All of the above

20: Which package is required to implement the machine learning model named conditional inference tree?

(a) Caret

(b) Lars

(c) Party

(d) None of the above

10

Evaluation and Validation of Results

The most common machine learning problems focused in this book are the classification and regression. These problems result in numerical prediction. There exist more than 200 machine learning models for both the learning problems. How to choose the appropriate model according to the problem requires consideration of two elements, i.e., experimental data and model. The following points should be considered before you perform various experiments.

1. Data:
    - The collected data should be structured.
    - The data should be noise-free and clean.
    - The size of data should be fit into the machine learning model.
    - The optimal features should be selected from the set of entire features.
    - The data is preprocessed in such a manner that it reduces the computational cost.
2. Machine learning model:
    - Machine learning model selection should be done according to the problem, i.e., supervised or unsupervised.
    - The focus should on analyzing of the working of each model to minimize the error rate.
    - Model should be selected by considering the minimum computational cost.
    - It should be analyzed that whether the model meets the goal of problem-solving or not.
    - It is important to analyze the parameters of the model such as how long it takes to compile the training dataset, and how scalable it is.

By considering all the above points, we are able to provide the preference for a few popular models. The comparison among the similar models is difficult. Like the old saying, when nothing is perfect, try all the alternatives. When the machine learning model results in minimum error using cross-validation, it seems to be the good model. Choosing the best one is again a question. This is not only a question but a challenge for researchers, who are working in the domain of machine learning.

The true performance of the model is the task performed with a statistical test by finding the confidence bounds. Experimenting with a particular dataset can help us to find confidence intervals but we can say that the coin tosses when the experimenting data is unlimited. Partitioning the training data can be one solution. If a limited number of samples are formed, and if there are enough samples, the mean of each sample represented by x, then the set x1, x2, x3...xn has a normal distribution. Once the task of machine learning model selection is done, then the performance of the selected model is evaluated. The performance of the model is evaluated using various parameters. The machine learning model is used to predict the unseen patterns. We make the system learn from the training dataset. Then, we test the system with the testing data. In other words, testing helps us to ensure how much the system has learned. Next, the predicted data is compared with the testing data. This comparison is drawn by various parameters. These evaluation parameters along with their statistics are discussed below.

10.1 Confusion Matrix

In the case of a two-class problem, where the target feature, or we can say the predicted feature consists of two possible values, we can test the predicted data against the testing data with the help of the confusion matrix. Here, a 2 by 2 matrix is generated. The diagonal elements represent the correctly classified elements, whereas the non-diagonal elements represent the misclassified elements. Table 10.1 shows the two-class confusion matrix and Table 10.2 shows the three-class confusion matrix.

TABLE 10.1: Confusion matrix: Two class

Class	**a**	**b**
a	0	1
b	1	0

TABLE 10.2: Confusion matrix: Three class

Class	**a**	**b**	**c**
a	0	1	1
b	1	0	1
c	1	1	0

The misclassified values are the cost incurred by the system for predicting the correctly classified values. The four important aspects of this matrix are defined as follows.

1. True positives (TP): The actual value is positive, and the predicted value is also positive.

2. True negatives (TN): The actual value is negative, and the predicted value is also negative.

3. False positives (FP): The actual value is negative, and the predicted value is positive.

4. False negatives (FN): The actual value is positive, and the predicted value is negative.

Let's understand thid with our previously used IoT dataset. We can implement the decision tree model with this dataset. Precipitation Type is the predicted variable. The value of this target feature is two-class, i.e., 0 or 60. The file is generated, which contains the actual and the predicted values. Here, the table shows the first 10 rows.

TABLE 10.3: Actual and predicted values

Row	**Actual**	**Predicted**
1	0	60
2	60	0
3	0	0
4	60	0
5	0	0
6	0	60
7	0	0
8	0	0
9	60	60
10	60	60

The confusion matrix for the data in given Table 10.3, is shown in Table 10.4.

TABLE 10.4: Confusion matrix

Class	**0**	**60**
0	TN=4	FP=2
60	TN=2	TP=2

10.2 Correlation

The variables of the dataset are related to each other in one form or the other. It is very important to understand the relationship among the variables. There can be the number of correlation among the variables such as one variable completely depends upon on the other, or the two variables depend upon the third variable, and so on. The value of this parameter mainly lies from -1 to 1. Depending upon this value, we can categorize the correlation into the following:

1. Strong correlation: If the value of correlation lies near to a positive one, then the variables are strongly correlated.
2. Negative correlation: If the value of correlation lies near the negative one, this means the variables are negatively correlated.
3. Weak correlation: If the value of correlation lies near to zero, this implies the weaker correlation.
4. No correlation: If the value of correlation is zero, this means there exists no relation at all.

10.2.1 Covariance

Covariance is the another type of relation among the variables, which is linear in nature. It is calculated from variation of the values from the center. The formula used for calculating variance is defined in Equation 10.1.

$$covariance(a, b) = (sum(a - mean(a)) * (b - mean(b))) * \frac{1}{n-1} \tag{10.1}$$

Now, with the help of covariance, we can easily find the correlation. There are two methods for computing correlation.

- Pearson's correlation
- Spearman's correlation

10.2.2 Pearson's Correlation

The Pearson correlation is used to represent the strength of correlation among the variables. The standard deviation along with the covariance is used for its computation as defined in Equation 10.2. As discussed above, the value of the correlation lies between -1 to 1.

$$Pearson's\_Correlation = \frac{covariance(a, b)}{(standard\_deviation(a) * standard\_deviation(b))} \tag{10.2}$$

We can calculate correlation while training the data in the decision tree as we did before. Just add Equation 10.3.

$$Pearson\_Correlation = \frac{covariance(Actual, Predicted)}{(sd(Actual) * sd(Predicted))} \tag{10.3}$$

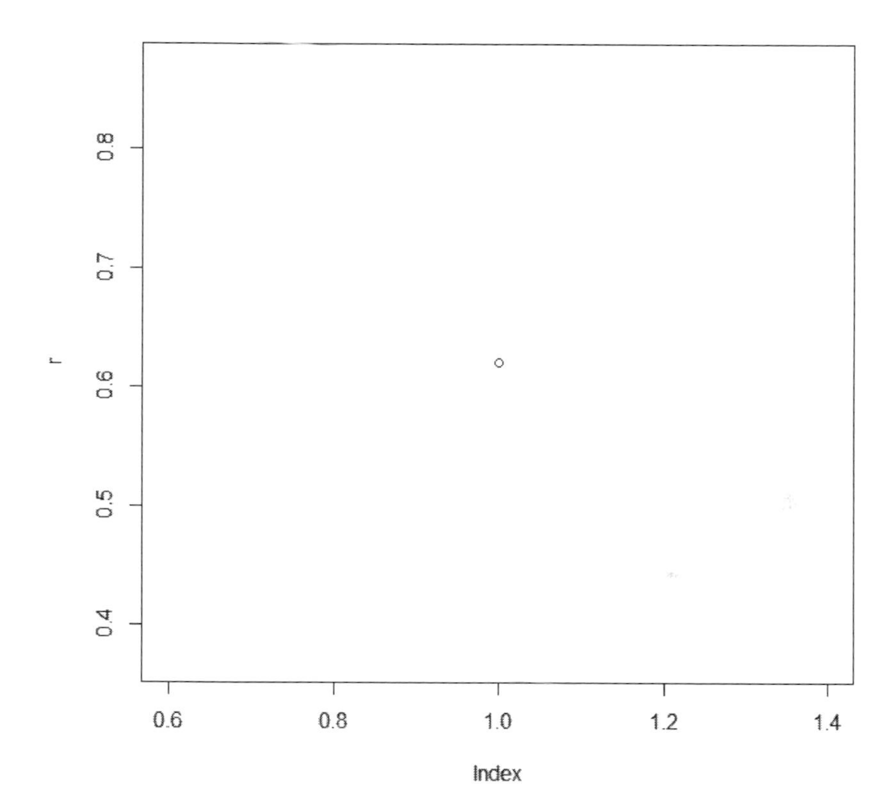

FIGURE 10.1: Pearson's correlation between the actual and the predicted

Here, in Equation 10.3 is the defined function in stats library of R, for computing standard deviation. This method is applied to the parametric data, i.e., data is having Gaussian distribution. We compute the correlation between the actual and predicted as shown in Figure 10.1.

10.2.3 Spearman's Correlation

This method is used to find the non-linear correlation between the non-parametric data, i.e., data may not have Gaussian distribution. This method is not applied directly to the data but on the rank of each sample according to Equation 10.4.

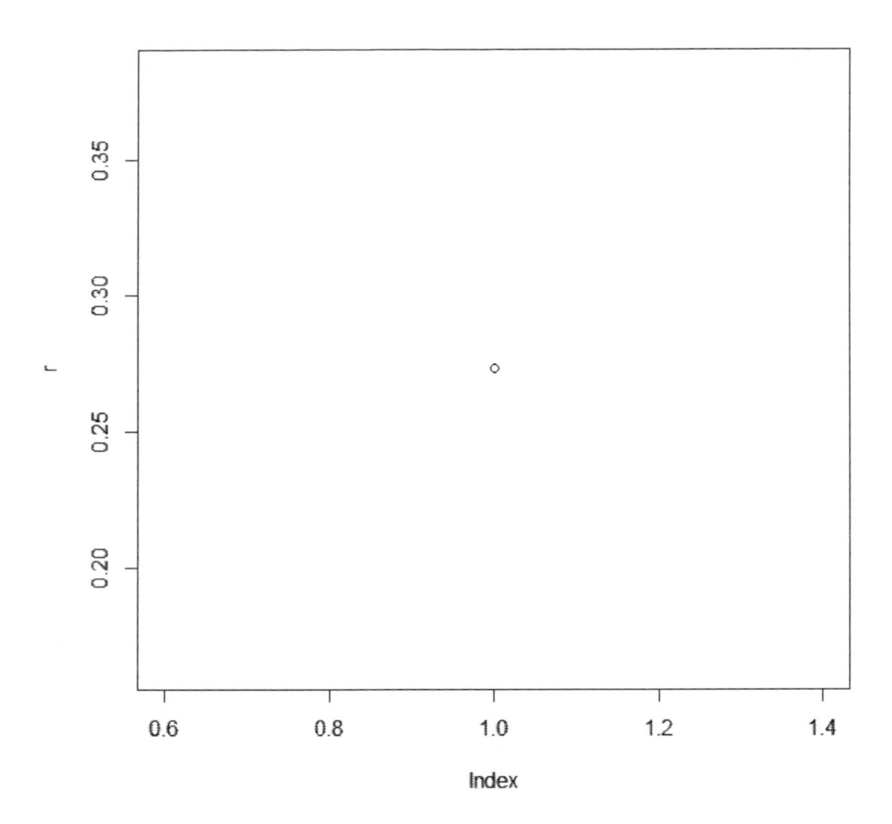

FIGURE 10.2: Spearman's correlation between the actual and the predicted

$$Spearman's\ correlation\ coefficient = \frac{covariance(rank(a), rank(b))}{standard\_deviation(rank(a))} \quad (10.4)$$

There is a default method in *R*, named 'Spearman' to find the Spearman correlation among the samples. The result is shown in Figure 10.2.

It can be computed without using the default Spearman method. The sample from the IoT dataset is considered for the illustration. Following are the steps to be followed while calculating the Spearman correlation.

1. Calculate the rank of each sample. The rank 1 is assigned to the highest value, 2 to the second-highest value and so on. The rank assignment is performed as shown in Table 10.5.

TABLE 10.5: Rank assignment

Humidity	**Wind Direction**	**Rank(humidity)**	**Rank(Wind Direction)**
55	63	3	8
56	124	2	7
54	156	4	4
53	150	5	6
52	155	6	5
58	175	1	1
58	164	1	2
45	155	7	5
32	159	9	3
42	155	8	5

2. Find the difference between the ranks and insert a new column in your data with name d. Also, compute the square of each rank. The difference between ranks and its square is computed and shown in Table 10.6.

TABLE 10.6: Rank difference computation: d(Rank(humidity))-(Rank(Wind Direction))

Rank(humidity)	**Rank(Wind Direction)**	**d**	**Square(d)**
3	8	5	25
2	7	5	25
4	4	0	0
5	6	1	1
6	5	1	1
1	1	0	0
1	2	1	1
7	5	2	4
9	3	6	36
8	5	3	9

3. Use this formula:

$$\rho = 1 - \frac{6 * \sum(d^2)}{n(n^2 - 1)} \tag{10.5}$$

When we compute ρ, 1- 6(102)/(10(100-1))
=> 1 - (612/990)
=> 0.38

10.2.4 Matthews' Correlation Coefficient (MCC)

MCC performs best in the case of binary classification. The values of MCC lies from -1 to 1. Each value signifies the type of prediction it holds. It is another way of describing the confusion matrix with a single term. It takes all the parameters of the confusion matrix into consideration as in Equation 10.6.

$$MCC = \frac{(TP * TN) - (FP * FN)}{\sqrt{(TP + FP)(TP + FN)(TN + FP)(TN + FN)}} \quad (10.6)$$

1. If MCC= +1, means it is a perfect prediction.
2. If MCC= -1, means its totally wrong prediction.
3. If MCC= 0, means it is a random prediction.

10.3 Coefficient of Determinant: R^2

It is computed in case of regression problems. This parameter states whether or not we can predict the one variable from another. In other words, the dependency of one variable over another. It is computed by taking the square of correlation. It takes the value from zero to one.

$$R^2 = correlation * correlation \quad (10.7)$$

The ratio of dependency is judged with the value of R^2 as discussed below:

- R^2= 0, means the dependent variable cannot be predicted from the independent variable.
- R^2= 1, means the dependent variable can be predicted without error from the independent variable.
- $0 < R^2 > 1$, means the extent to which the dependent variable is predictable. If R^2= 0.90, this means the 90% of the variance in a is predictable from b.

10.4 Accuracy (ACC)

Accuracy is the percentage difference between the predicted target and actual target. In other words, how many values are correctly predicted out of the total values. The Equation 10.8 is used for computing accuracy.

$$Accuracy = \frac{(TP + TN)}{(TP + FP + FN + TN)} \quad (10.8)$$

Let's compute the accuracy of the data given in Table 10.3. According to the data, we compute the parameters:

- TP= 4
- TN= 2
- FP= 2
- FN= 2

Substituting these parameters in Equation 10.8,

$$\text{Accuracy} = (4+2)/10 = 0.6$$

For computing the percent deviation, it figures out to be 60%. Accuracy does not give you a clear picture of the model's performance. It only tells about the correct predictions.

10.5 ROC and AUC

ROC is the receiver operating characteristic curve which shows the performance of the classification model. AUC is the area under the curve of ROC. That is why it is also known as AUROC (area under the receiver operating characteristics). It is formed with the help of two parameters, i.e., TPR and FPR. These parameters are discussed in the next section. The layout of these curves is as shown in Figure 10.3.

10.6 Error in Regression

The values in the regression problem vary within range but are not defined as absolute values. The same is true of the errors.

10.6.1 Root Mean Squared Error (RMSE)

It ranges from 0 to infinity. It is a rule which is quadratic based, and it measures the average magnitude of the error. In other terms, we can say it is the difference between the actual and the predicted and squaring it, followed by a mean and square root. It can be used for the comparison of models that produces the errors in same units.

$$RMSE = \sqrt{\frac{\sum_{i=1}^{n}(p_i - a_i)^2}{n}} \tag{10.9}$$

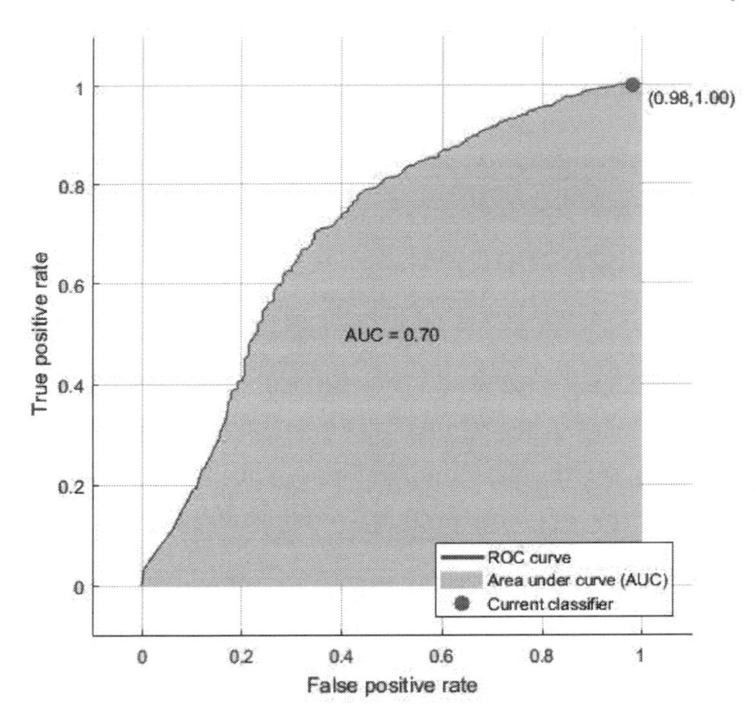

FIGURE 10.3: ROC and AUC

10.6.2 Mean Absolute Error (MAE)

It is similar to RMSE, but relatively small. It does not take the square of the difference of actual and predicted but instead takes their absolute values. It also lies between 0 and infinity.

$$MAE = \frac{\sum_{i=1}^{n} |p_i - a_i|}{n} \tag{10.10}$$

10.6.3 Relative Squared Error (RSE)

It can be used as the parameter of comparison for the models, which produces the error in different units.

$$RSE = \sqrt{\frac{\sum_{i=1}^{n} (p_i - a_i)^2}{\sum_{i=1}^{n} (\bar{a} - a_i)^2}} \tag{10.11}$$

10.6.4 Relative Absolute Error (RAE)

It can be used for the comparison of models, those produces the errors in different units.

$$RAE = \sqrt{\frac{\sum_{i=1}^{n} |p_i - a_i|}{\sum_{i=1}^{n} |\bar{a} - a_i|}} \tag{10.12}$$

10.7 Measuring Rates

10.7.1 Sensitivity, Recall, Hit Rate, or True Positive Rate (TPR)

$$TPR = \frac{TP}{TP + FN} \tag{10.13}$$

10.7.2 Specificity, Selectivity, or True Negative Rate (TNR)

$$TNR = \frac{TN}{TN + FP} \tag{10.14}$$

10.7.3 Precision, Positive Predictive Value (PPV)

$$PPV = \frac{TP}{TP + FP} \tag{10.15}$$

10.7.4 Recall, Sensitivity, Hit Rate, or True Positive Rate (TPR)

$$TPR = \frac{TP}{TP + FN} \tag{10.16}$$

10.7.5 Fallout, False Positive Rate (FPR)

$$FPR = \frac{FP}{FP + TN} \tag{10.17}$$

10.7.6 Miss Rate or False Negative Rate (FNR)

$$FNR = \frac{FN}{FN + TP} \tag{10.18}$$

10.7.7 False Discovery Rate (FDR)

$$FDR = \frac{FP}{FP + TP} \tag{10.19}$$

10.7.8 False Omission Rate (FOR)

$$FOR = \frac{FN}{FN + TN} \tag{10.20}$$

10.8 F Measure

It is the ratio of precision and recall. Other measures, such as accuracy, do not consider false positives and false negatives. These estimates are covered in F measure. In other words, we can say, it is the harmonic mean between sensitivity and precision. The value of F lies between 0 and 1. F measure can be computed with multiple equations using Eq. 10.21 to Eq. 10.25.

$$F = 2 * \frac{(PPV * TPR)}{(PPV + TPR)} \tag{10.21}$$

$$F = \frac{2TP}{2TP + FP + FN} \tag{10.22}$$

$$F = \frac{1}{\frac{\frac{1}{Recall} + \frac{1}{Precision}}{2}} \tag{10.23}$$

$$F = 2 * \frac{Precision * Recall}{Precision + Recall} \tag{10.24}$$

$$\tag{10.25}$$

10.9 Summary and What's Next?

Exploring the functionalities of machine learning models is not enough until we learn to deploy the most suitable model for the problem. In this chapter, we learned what the various constraints are to keep in mind while choosing the machine learning model. We also covered the various evaluation parameters, for both classification problems and regression problems. These parameters are the confusion matrix, accuracy, R^2, F measure, recall, precision, TPR, TNR, FNR, FPR, and different kinds of errors. So evaluating measures for detecting the performance of machine learning models can be done in an efficient manner.

10.10 Exercises

Questions

1. What are the different parameters considered before performing experiments?
2. What are the different parameters for the selection of a machine learning model?
3. Explain confusion matrix with an example.
4. What is correlation? What are its various categories?
5. Define covariance.
6. Explain Matthews correlation coefficient in detail.
7. What is the need for computing a coefficient of determinant?
8. State all the parameters computed for determining the success of a classification machine learning model.
9. State all the parameters computed for determining the success of a regression machine learning model.
10. What are the different types of error rates computed in the problem of regression?
11. Compute all possible evaluation parameters for the data in Table 10.7:

TABLE 10.7: Error matrix

Class	**Actual**	**Predicted**
1	1	0
2	0	1
3	1	1
4	0	0
5	1	1
6	1	0
7	0	0
8	0	0
9	1	1
10	0	0
11	1	1
12	0	1
13	1	1
14	0	0
15	1	1
16	0	0
17	1	1
18	0	1
19	1	0
20	0	0

Multiple Choice Questions

1: Which aspect is considered important before performing experiments?

(a) Data structure

(b) Data features

(c) Data size

(d) All of above

2: Which aspect is considered important for selecting a of machine learning model?

(a) Problem type

(b) Error rate

(c) Computational cost

(d) All of the above

3: The diagonal elements in the confusion matrix represent:

(a) Correctly classified elements

(b) Wrongly classified elements

(c) Error

(d) None of the above

4: The non-diagonal elements in the confusion matrix represent:

(a) Correctly classified elements

(b) Wrongly classified elements

(c) Error

(d) None of the above

5: Which parameter represents the actual value as positive and the predicted value as positive?

(a) True Positives (TP)

(b) True Negatives (TN)

(c) False Positives (FP)

(d) False Negatives (FN)

6: Which parameter represents the actual value as negative and the predicted value as negative?

(a) True Positives (TP)

(b) True Negatives (TN)

(c) False Positives (FP)

(d) False Negatives (FN)

7: Which parameter represents the actual value as negative and the predicted value as positive?

(a) True Positives (TP)

(b) True Negatives (TN)

(c) False Positives (FP)

(d) False Negatives (FN)

8: Which parameter represents the actual value as positive and the predicted value as negative?

(a) True Positives (TP)
(b) True Negatives (TN)
(c) False Positives (FP)
(d) False Negatives (FN)

9: What kind of correlation is it, if it lies near to positive one?

(a) Strong correlation
(b) Negative correlation
(c) Weak correlation
(d) No correlation

10: Which correlation is used to find the linear and strong correlation among the data?

(a) Spearman's correlation
(b) Pearson's correlation
(c) Strong correlation
(d) None of the above

11: Matthews' correlation coefficient is adopted in case of:

(a) Binary classification
(b) Multi-classification
(c) Regression
(d) Clustering

12: In which kind of machine learning problem, the coefficient of determinant is computed?

(a) Binary classification
(b) Multi classification
(c) Regression
(d) Clustering

13: AUC is computed using:

(a) TPR
(b) FPR
(c) a & b both
(d) None of the above

14: Which parameter is used to determine the average magnitude of the error?

(a) RMSE
(b) MAE
(c) RSE
(d) RAE

15: Which parameter returns the exact mean of error?

(a) RMSE
(b) MAE
(c) RSE
(d) RAE

16: What are the different parameters considered for computing F measure?

(a) Precision & Recall
(b) Precision & Sensitivity
(c) Recall & Sensitivity
(d) None of the above

17: What kind of correlation is it, if it lies near to negative one?

(a) Strong correlation
(b) Negative correlation
(c) Weak correlation
(d) No correlation

18: What kind of correlation is it, if it lies near to zero?

(a) Strong correlation
(b) Negative correlation
(c) Weak correlation
(d) No correlation

19: What kind of correlation is it, if its value is zero?

(a) Strong correlation
(b) Negative correlation
(c) Weak correlation
(d) No correlation

20: How many instances are formed in a three-class confusion matrix?

(a) 6
(b) 7
(c) 8
(d) 9

11

Solutions

11.1 Chapter 1

1. C
2. B
3. B
4. A
5. B
6. B
7. A
8. B
9. B
10. A
11. C
12. C
13. B
14. C
15. A
16. A
17. D
18. B
19. B
20. B

11.2 Chapter 2

1. A
2. D
3. B
4. B
5. B
6. D

7. A
8. C
9. A
10. A
11. A
12. D
13. B
14. D
15. B
16. B
17. C
18. C
19. B
20. B

11.3 Chapter 3

1. D
2. A
3. B
4. C
5. D
6. B
7. C
8. C
9. C
10. D
11. D
12. A
13. C
14. B
15. D
16. A
17. C
18. A
19. D
20. D

11.4 Chapter 4

1. D
2. C
3. C
4. B
5. A
6. B
7. D
8. B
9. B
10. D
11. B
12. B
13. D
14. D
15. C
16. B
17. A
18. A
19. A
20. C

11.5 Chapter 5

1. A
2. B
3. D
4. B
5. D
6. C
7. C
8. B
9. D
10. C
11. C
12. C
13. C
14. B
15. B

16. B
17. D
18. A
19. C
20. B

11.6 Chapter 6

1. A
2. B
3. C
4. B
5. A
6. B
7. A
8. A
9. C
10. A
11. B
12. A
13. B
14. A
15. C
16. A
17. B
18. B
19. A
20. B

11.7 Chapter 7

1. C
2. B
3. B
4. B
5. A
6. B

7. B
8. A
9. B
10. B
11. D
12. A
13. D
14. D
15. D

11.8 Chapter 8

1. B
2. B
3. A
4. D
5. C
6. A
7. A
8. A
9. B
10. A
11. A
12. C
13. B

11.9 Chapter 9

1. A
2. D
3. D
4. C
5. D
6. D
7. C
8. A
9. B

10. D
11. A
12. C
13. A
14. B
15. B
16. A
17. A
18. B
19. B
20. C

11.10 Chapter 10

1. D
2. D
3. A
4. D
5. A
6. B
7. C
8. D
9. A
10. B
11. A
12. C
13. C
14. A
15. B
16. A
17. B
18. C
19. D
20. D

12

Dataset

The complete details of the IoT dataset being used for experiments is discussed in Section 4.2.1. Mentioning the complete dataset is out of the scope of this book. The dataset contain 59145 rows in total, and 16 columns (2 columns already because of the absence of data). We have applied a filter on the column rain intensity to lower the zero values for size compression. The refined data now consists of 1519 rows as mentioned below. The parameters such as station name, are removed of their data type and the rest of the column names are:

- A: Wet Bulb Temperature
- B: Humidity
- C: Rain Intensity
- D: Interval Rain
- E: Total Rain
- F: Precipitation Type
- G: Wind Direction
- H: Wind Speed
- I: Maximum Wind Speed
- J: Solar Radiation
- K: Heading
- L: Battery Life

A	B	C	D	E	F	G	H	I	J	K	L
14.3	73	1.2	0.4	3	60	147	1	3.8	40	322	12
13.4	82	7.2	5.8	8.9	60	47	1	1.4	47	322	12.1
19.7	86	2.4	0.3	9.6	60	356	1.3	2.4	58	0	12.1
18.8	80	0.6	0.7	10.3	60	344	1.2	2.4	52	0	12.1
18.5	89	0.6	0.5	10.8	60	354	0.7	1.1	68	0	12.1
8.6	93	0.6	1.8	14.6	60	6	8.2	12.1	51	359	12.1

A	B	C	D	E	F	G	H	I	J	K	L
8.6	93	0.6	1.8	14.6	60	6	8.2	12.1	16	359	12.1
8.2	95	1.8	2.6	2.6	60	12	7.7	12.3	26	359	12.2
7.5	92	5.4	1.1	3.7	60	16	8.8	12.5	12	359	12
7.4	91	1.2	1.8	5.5	60	3	10.2	13.3	5	359	12.1
7.5	91	0.6	3.3	8.9	60	26	9	12.7	3	359	12.1
17.8	89	4.8	4.9	15.2	60	254	1.8	6.2	20	0	12.1
17.4	74	10.8	4.2	23.6	60	266	2.2	4.6	10	0	12.1
18.3	73	11.4	2.3	26.1	60	281	1.4	3.6	8	0	12.1
19.4	92	1.8	0.4	33.2	60	321	1.2	3.4	26	0	12.1
19.1	93	6	0.5	33.7	60	340	1.3	2.3	47	0	12.1
12.8	95	8.4	0.5	34.7	60	33	2.7	3.3	30	359	12.1
18.6	86	8.4	2.7	37.8	60	324	2.1	3.1	2	359	12.1
20.4	80	4.2	2.6	40.4	60	297	1.4	3.8	0	0	12.1
20.6	93	9.6	4.6	46.5	60	17	1	1.7	0	0	12.1
21	89	3	0.2	52.3	60	289	2.1	5.8	3	0	12.1
20.8	94	5.4	9.3	61.5	60	216	0.4	3	4	0	12.1
21.3	95	9	4.4	66.7	60	266	0.9	2.5	28	359	12.1
21	96	0.6	1.6	70.6	60	72	0.4	0.7	98	0	12.1
22.2	92	0.6	4.3	74.9	60	246	0.5	1.8	32	0	12.1
21.6	88	13.8	1.1	77.8	60	28	1.6	2.4	6	0	12.1
20.3	93	12	10.7	88.5	60	159	0.6	1.6	7	0	12.1
20.5	95	4.8	9.5	98	60	123	0.6	1.2	2	0	12.1
21.1	94	13.8	8	106	60	263	1.3	2.9	1	0	12.2
21.3	92	1.2	3.5	109.5	60	337	1	2.3	0	0	12.1
5.9	86	7.2	5	5.2	60	119	5.1	7.1	38	354	12
13.1	92	1.2	0.4	112.2	60	30	1.3	1.9	238	0	12.1
19.6	93	3.6	5.3	123.1	60	168	0	1.2	0	0	12.1
24.3	87	2.4	0.5	208.2	60	258	4.5	6.7	22	355	12
23.2	89	0.6	3.9	212.1	60	73	0.7	1.4	164	355	11.8
25.5	77	4.8	0.1	212.2	60	255	6.3	10.9	0	355	11.9
19.3	86	6.6	2.3	214.6	60	241	2.7	4.9	1	354	12
18.9	76	3	0.9	130	60	343	0.6	2.8	0	0	12.1
18.9	95	4.8	7.4	222	60	15	2.6	4.1	2	355	12
18.8	93	4.8	6.8	136.8	60	0	0	0.5	0	0	12.2
18.9	97	12.6	8	230	60	198	0.8	1.6	2	355	12
19.1	91	5.4	3.1	140	60	275	1.3	2.6	0	359	12.1
16.8	97	10.2	4.4	240.5	60	28	4	5.9	46	355	11.9
16.4	95	6.6	4.8	245.3	60	4	4.4	7.8	50	355	11.9
15.5	89	5.4	0.3	146.2	60	0	0.5	1.9	21	0	12.1
16.4	97	3.6	6.6	251.9	60	10	5.4	8.2	24	354	11.9
15.3	91	0.6	3.2	149.4	60	10	3.1	4.5	28	0	12.1
15.8	96	1.2	9.8	261.7	60	8	4.4	7.8	27	354	11.9
14.9	92	2.4	0.9	150.3	60	37	2	3	27	359	12

A	B	C	D	E	F	G	H	I	J	K	L
16	96	4.8	5.9	267.6	60	26	4.8	8.5	11	354	11.9
15.7	95	4.8	7	274.6	60	16	6.1	8.8	5	354	11.9
14.8	91	1.8	1.1	154.8	60	359	1.3	1.7	5	359	12
15.8	94	1.2	0.6	280.5	60	354	3.2	4.7	2	354	11.7
18.2	64	0.6	0	280.9	60	206	1.6	2.9	95	355	12
18.2	65	0.6	0.1	281	60	239	1.8	3	45	354	12
18.3	91	1.8	1.6	158.8	60	249	1.2	3.5	1	0	12.1
21.9	96	0.6	0.4	317.6	60	208	1.6	2.3	1	355	12
21.9	66	4.8	0.4	162.3	60	347	2.7	4.5	72	0	12.2
22.2	92	5.4	5.7	310.1	60	271	4	7.2	1	355	11.9
21.5	91	10.8	7.5	174.1	60	261	3.1	7.8	0	0	12.1
22.1	95	0.6	3.1	315.8	60	276	0.7	2.3	1	355	12
21.3	89	4.2	3	177.5	60	328	1.2	2.7	0	0	12.1
21.3	90	1.2	3.6	181.1	60	218	0.5	1.6	0	0	12.1
21.9	95	5.4	0.5	318.1	60	225	2.3	3.8	1	355	12
21.4	88	0.6	0.1	181.2	60	0	0.4	1.6	0	0	12.1
21.4	87	1.2	0.1	181.3	60	271	1	3	0	0	12.1
15.9	97	1.8	3.6	322.5	60	8	8.1	15.6	2	354	12
15.2	92	0.6	1.5	182.8	60	358	5.8	9.3	2	359	12.1
14.7	97	0.6	1	329.2	60	14	9.9	16.9	8	354	11.9
14.3	92	3.6	0.9	184	60	9	7.5	9.7	7	359	12.1
17.1	97	0.6	0.3	330.5	60	77	6.3	7.6	2	355	11.9
20.1	80	0.6	0.7	331.8	60	162	1.3	2.7	99	355	12
19.7	85	1.2	0.5	331.1	60	163	1.1	2	101	355	12
19	71	1.2	0.3	185	60	116	1.4	3.1	161	0	12.1
19.8	83	0.6	0	185.5	60	105	1.1	2.6	0	0	12.1
21.1	87	1.8	0	332.4	60	282	4.5	13.6	1	354	12
20.3	82	7.8	1.4	187	60	4	4.6	10.9	0	0	12.1
22.7	72	0.6	0.4	342.5	60	243	4.1	5.9	230	355	12
20.5	96	0.6	0.4	376.3	60	169	1.3	2.5	1	354	11.9
20.3	83	3	0.4	362.4	60	68	0.5	1.5	20	354	12
18.1	85	2.4	0.3	191.9	60	81	0.7	1.4	10	0	12
18.8	96	12.6	10.7	373.1	60	101	2	4.2	7	355	12
18.5	88	11.4	8	200	60	118	1.5	3.2	12	0	12.1
19.7	95	1.2	5.5	208.1	60	111	2	4.2	0	0	12.1
26.1	76	31.2	4.9	394	60	275	4.5	8.3	30	355	12
19.8	80	9.6	3.1	431.8	60	21	8.1	14.7	1	355	11.9
22.9	46	1.8	0	220.4	60	263	1.9	6.1	10	1	12.2
19	99	1.8	22.3	454	60	216	1.7	3.9	2	355	11.9
20.7	84	5.4	0.5	457	60	8	4.8	8.7	29	355	12
20.2	73	0.6	0.1	228.3	60	354	1.5	3.3	41	0	12.1
19.7	90	18.6	8.7	467	60	8	4.2	8.5	2	354	12
19.7	84	8.4	4.3	232.6	60	355	2.4	5.9	1	0	12.1

A	B	C	D	E	F	G	H	I	J	K	L
20.4	91	3.6	8.6	475.6	60	247	3.5	4.8	1	354	12
19.7	88	19.2	2.9	481.3	60	340	3.4	6.2	1	354	12
19.2	86	6	10.1	249.7	60	40	1.3	1.7	1	0	12.1
20.3	90	14.4	4.3	239.7	60	292	0.8	1.5	0	0	12.1
19.8	89	4.8	3.9	253.7	60	268	1.1	4.9	1	0	12.1
20.5	86	3	1.6	258.9	60	262	2.3	5.4	0	0	12.1
20.5	87	1.2	6.9	265.8	60	267	0.6	1.4	1	0	12.1
20.6	97	22.2	4.4	494.7	60	209	2.7	5.5	1	355	12
20.3	88	18.6	1.8	267.6	60	269	1.3	2.7	1	0	12.1
20.6	97	7.8	11.8	506.6	60	176	3.9	5.3	1	355	12
19.8	88	0.6	7.6	275.2	60	0	0	0.4	1	0	12.1
20.1	96	1.8	4.4	511	60	166	2.8	4.1	2	354	11.9
20.1	92	0.6	0.8	276	60	116	1	2.8	1	0	12.1
14.9	64	1.2	0	284.9	60	262	2.3	6.7	1	0	12.1
23.3	78	0.6	0	511.4	60	172	2.7	6.3	11	355	11.9
23.6	81	2.4	0.8	512.2	60	174	3	5	0	355	12
22.9	79	0.6	0	276.1	60	93	1.4	3.3	1	0	12.1
20.1	99	2.4	24.6	536.8	60	235	2.4	3.7	2	354	12
19.8	94	3.6	7.7	283.8	60	316	1.1	1.7	1	0	12.1
17.2	89	2.4	6.7	297.5	60	124	0.8	1.4	1	0	12
17.3	74	0.6	0.1	285.3	60	115	1.2	2.8	0	359	12
17.3	87	3.6	2.1	557.6	60	302	0.7	1	2	354	11.8
17.1	83	8.4	5.5	290.7	60	115	0.8	1.8	1	0	12.1
17.2	90	5.4	1.5	561.5	60	160	1.2	2.3	6	353	11.9
17.3	93	1.8	2.6	564.1	60	0	0	0.8	28	354	11.9
18.5	90	0.6	0.1	305.1	60	34	0.6	1.4	49	0	12
20.9	97	83.4	32.3	597.5	60	85	2.5	3.1	6	354	12
23.1	78	15.6	0.9	308.1	60	16	0.6	1.3	0	0	12.1
21.7	71	7.8	0.2	308.5	60	344	3.1	6.8	0	0	12.1
22.3	100	6.6	39.7	640.1	60	247	4.6	9.4	5	355	11.9
21.9	91	3.6	7	315.5	60	92	0.7	1.9	8	0	12.2
22.6	96	3.6	1.3	648.4	60	243	3	5.5	28	355	12
22	86	0.6	2.2	321.6	60	294	0.9	2.2	26	0	12.1
21.1	83	0.6	0.3	322.4	60	0	0.4	0.7	0	0	12.1
16.3	70	11.4	0.6	323.1	60	338	2.2	2.9	37	0	12.2
17	89	2.4	3.3	659.6	60	203	2	4.4	37	355	12
16.3	72	0.6	7.6	330.7	60	99	0.5	1.4	28	0	12
16.5	84	1.2	8.2	667.8	60	182	2.4	5	72	354	11.9
16.3	94	1.2	0.6	669.6	60	318	1.5	4	45	354	11.9
16	89	1.2	1	333.1	60	335	1.3	2.2	30	0	12
14.9	82	4.8	2.3	672.8	60	70	6.6	8.7	92	355	11.9
15.5	83	1.2	1.4	335.4	60	27	6.2	8.2	54	0	12
15.3	77	3.6	0.3	336.8	60	24	5.9	6.7	45	0	12.1

A	B	C	D	E	F	G	H	I	J	K	L
11.6	95	9.6	10.5	690.9	60	323	2.4	5.2	3	354	11.9
11.4	83	2.4	1.7	338.6	60	284	1.5	4.7	3	359	12
12.4	84	7.8	8.4	360.3	60	353	3.8	6.6	29	0	12.1
12.8	85	13.8	4.5	704.2	60	16	9.3	13.1	3	354	11.9
12.2	88	16.8	11	715.2	60	40	6.2	11.6	6	354	12
11.8	88	9	6	351.8	60	21	5.3	7.5	6	359	12.1
12.1	79	3	0.3	363.5	60	4	5.2	9.5	56	0	12
11.9	86	0.6	4.3	744.2	60	349	7.7	11.3	80	354	11.9
11.4	78	1.2	0.6	364.1	60	347	4.4	7.8	145	0	12.1
11.3	81	21	7.5	751.7	60	347	5.9	10	86	354	11.9
11.3	72	1.2	0.8	364.9	60	354	3.7	6.3	249	0	12.1
19	95	14.4	6	375	60	22	3.7	5.3	1	0	12.1
19.9	99	22.8	27.6	789.7	60	50	0.8	1.9	1	354	11.9
19.6	96	24.6	8.8	383.7	60	350	0.7	1.8	1	0	12.2
20.5	100	1.8	21.2	811	60	212	2	3.3	1	354	12
20	94	1.2	4.4	388.2	60	100	0.6	0.9	0	0	12.1
20.5	100	1.2	4.6	815.6	60	172	3.4	5.1	1	355	11.9
20.1	95	4.2	4	392.2	60	0	0.7	2.2	0	0	12.1
20	96	1.2	6.7	857.5	60	0	0	0.8	1	355	11.9
20.2	90	0.6	0.4	395.8	60	120	1.2	2.4	0	0	12
20.3	98	0.6	0.1	820.3	60	79	1.3	1.7	2	354	12
20.9	97	13.8	30.6	850.8	60	169	3.2	4.7	2	354	11.9
20.2	86	1.8	2.2	395.4	60	118	0.7	1.3	4	0	12.1
20.5	98	4.2	3.4	861	60	80	1.3	1.5	1	354	11.9
19.5	95	4.8	6.1	402	60	18	1.3	2	0	0	12.1
21.1	99	1.2	3.3	864.2	60	186	1.6	2.4	1	355	11.9
18.5	100	0.6	14.8	879.1	60	155	0.7	1.7	2	354	11.9
15.4	77	1.2	0	879.9	60	19	9.2	14	51	354	12
8.3	67	0.6	0.5	1.4	60	74	7.8	11.3	3	0	12
13	76	4.2	4	8.8	60	296	1.4	5.7	1	0	12.1
12.1	71	0.6	0.1	6.2	60	144	2.2	4.4	101	354	11.9
12.8	64	0.6	0.6	6.9	60	168	4.1	6.4	2	354	11.9
13.3	63	0.6	0	7.8	60	168	3.6	6.3	2	354	11.9
18.1	88	0.6	1	12.5	60	283	2.2	4.9	43	0	12.1
18.3	87	3.6	0.7	13.2	60	259	2.9	12.6	21	0	12.1
9.4	67	1.8	0.4	10.9	60	125	4.5	6.1	76	354	11.9
10.3	69	1.2	3.8	15.5	60	109	5.8	8.3	163	354	11.9
11.5	95	7.8	1.2	53.4	60	135	7.5	10	3	354	11.9
11.1	91	3	0.3	3.2	60	104	2.3	6	1	0	12.2
12.1	94	1.8	1	56.4	60	167	4.2	7	3	354	11.9
13.5	97	3.6	1.8	60.3	60	219	3.2	6.6	46	354	11.9
9.7	67	1.2	0.6	11.7	60	107	5.4	7.7	150	353	11.9
10	64	0.6	0.1	14.7	60	106	3.7	5.3	164	0	12

A	B	C	D	E	F	G	H	I	J	K	L
10.8	79	3.6	3.6	21.9	60	131	7.3	11.6	3	354	11.9
10.6	83	2.4	13.3	35.2	60	127	8	11.8	3	354	11.9
10.6	82	2.4	3.2	21	60	106	2.4	9.3	1	0	12.1
10.8	85	4.2	8.7	43.9	60	123	8.2	10.7	3	354	12
10.6	82	2.4	3.2	21	60	106	2.4	9.3	1	0	12
10.4	88	3	2.7	2.7	60	314	1.3	7	1	0	11.9
12.1	96	0.6	0.8	55	60	149	3.1	6.7	3	354	11.9
11.6	93	1.2	1.1	5.6	60	106	1.9	5.3	1	0	12.1
7	86	1.2	1	66.1	60	177	3.7	6.8	23	354	11.9
7.2	79	0.6	0.7	9.6	60	76	2.3	5	22	0	12.1
9	95	5.4	6.2	96	60	199	4	6.7	19	354	11.9
7.1	89	4.2	1.5	67.6	60	188	3.8	7	54	354	11.9
7	82	2.4	1.1	65	60	167	3.5	6.3	13	354	11.9
7.1	77	0.6	1.1	8.9	60	104	2.8	4.3	10	0	12
7.3	83	1.2	0.7	10.3	60	106	3.7	6.4	38	0	12.1
7.2	93	12	8.8	76.4	60	185	4.4	7.7	36	354	11.9
7.5	88	4.2	6	16.3	60	309	1	3.6	34	0	12
7.5	94	6	3.7	80.1	60	164	3.7	6.6	47	354	12
7.6	90	6	2.4	18.7	60	106	3.6	7	50	0	12.1
7.7	93	4.8	8	88.1	60	173	3.6	6.9	63	354	11.9
8.3	90	1.2	3.9	22.6	60	118	3.2	5.7	46	0	12
17.2	75	0.6	0.1	26.7	60	276	3.6	10.5	-2	0	12.1
17.5	85	3.6	0.4	27.1	60	279	2.5	5.8	0	0	12
17.9	91	2.4	6.2	9.9	60	216	5.4	9.6	2	355	12
17.5	91	2.4	5.5	32.7	60	280	2.1	3.3	0	0	12.1
11.2	63	1.8	0.6	39.3	60	106	3.2	10.7	0	0	12.1
9.8	90	4.8	1.3	1.3	60	272	3.3	9.8	1	0	12.1
8.6	82	0.6	0	3.3	60	113	3.9	6.9	1	0	12
10	81	0.6	0	0.5	60	169	4.5	8	3	354	11.9
10.1	81	0.6	0.1	4.6	60	100	2.2	5	8	0	12.1
12.5	77	2.4	1.1	2	60	167	6	10.2	73	354	11.9
12.8	82	24.6	6.2	8.2	60	162	6.6	10.6	49	354	12
11.9	86	0.6	2.6	7.4	60	112	2.9	5.6	27	0	12
12.8	87	12.6	9.9	18.2	60	165	7.6	15.4	22	354	11.9
12.4	85	15	4.1	11.5	60	84	2	4.6	5	0	12
12.3	92	8.4	9.8	21.3	60	101	3.7	5.9	3	0	12
12.4	71	1.8	4.2	34.3	60	189	6.5	10.4	3	354	11.9
12.4	67	1.2	1	22.8	60	62	2.3	5.1	0	0	12.1
13.4	78	3.6	1.7	36.9	60	175	8.3	12	2	354	11.9
13	73	3	0.4	24.2	60	11	2.1	3.8	0	0	12.1
13.5	84	1.8	5.9	42.8	60	174	7	12.5	3	354	11.9
14.8	92	0.6	1.1	45.2	60	207	5.1	9.2	14	355	11.9
14.5	84	0.6	0.4	26.1	60	85	1.2	2.6	8	0	12

A	B	C	D	E	F	G	H	I	J	K	L
2	97	4.8	3	73.5	70	103	9.4	13.5	6	354	11.9
2.6	89	1.8	1.1	47.3	60	106	6	9.2	4	0	12.1
1.3	93	2.4	2.5	49.8	60	13	3.3	5.7	4	353	11.9
0.7	90	3.6	3.3	30.1	60	339	0.5	1.3	3	0	12.1
1.7	88	0.6	1	38.3	60	72	4	4.9	3	0	12.1
1.7	98	1.8	6.6	84.7	60	89	8.8	11.7	29	354	11.9
2.9	91	1.2	3	52.1	60	90	6.7	10	28	0	12.1
1.5	97	4.2	3.4	66.4	60	109	8.1	10	4	353	11.9
2.3	89	2.4	1.3	42.9	60	97	5.9	9	3	359	12
0.8	79	2.4	0.3	26.8	60	35	1.2	1.5	3	0	12.1
1.7	97	4.8	4.1	70.5	60	96	8.4	11	4	354	11.9
1.8	91	2.4	3.2	46.1	60	112	5.6	8.4	3	0	12.1
1.3	96	3	2.2	52	60	14	5	7.3	4	354	11.9
1.7	96	3	2.5	54.6	60	57	3.8	6.6	4	354	11.9
1.2	92	3.6	3.4	37.3	60	43	1.9	2.7	3	0	12.1
1.6	96	3	1.4	57	60	91	5.2	8.2	4	354	11.9
1.8	88	0.6	1.1	39.3	60	101	5.6	7.6	3	0	12.1
1.9	96	0.6	1	58	70	94	6.4	9	4	353	11.9
2.1	88	0.6	0.9	40.2	60	101	4.8	6.6	3	0	12
1.8	96	2.4	2	60	60	96	6.7	9.4	4	353	11.9
1.9	97	3.6	3	63	70	99	7.7	11.6	4	354	11.9
2.1	87	0.6	1.2	41.6	60	108	4.6	7.4	2	0	12
2.2	98	3.6	4.6	78.2	70	92	7.7	11.7	9	354	11.9
2.5	89	2.4	1.8	49.1	60	100	6.6	10	12	359	12
1.9	98	7.2	3.7	88.4	70	88	8.3	12.6	27	354	11.9
1.8	85	1.2	1	53	60	91	6.5	8.8	29	0	12.1
2	98	3.6	0.8	0.8	70	77	11.6	14	52	354	11.9
2.1	90	0.6	2.1	55.1	60	91	7.2	9.2	47	0	12.1
2.1	98	2.4	3.4	3.4	60	92	8.2	12.4	65	354	11.9
2.2	87	0.6	0.6	55.7	70	80	6.4	8	39	0	12.1
1.6	98	3.6	3.4	6.9	70	86	5.1	6.7	32	354	11.9
1.7	90	1.2	1.7	57.4	60	73	3	4.7	21	0	12.1
0.9	98	2.4	1.7	8.6	60	323	1.9	5	19	354	11.9
1.8	94	0.6	1.7	59.1	60	23	7.3	9.7	16	0	12.1
0.3	98	1.8	2.7	11.3	60	332	4.3	8	18	354	11.9
0.2	91	2.4	1.8	61	70	12	2.6	5.5	15	0	12
9.9	86	1.2	2.4	4	60	214	4.4	8.9	3	354	11.9
9.3	81	1.2	0.5	1.5	60	223	3.4	5.7	3	354	12
9.9	84	1.8	1.2	63.9	60	283	2.8	7	1	1	12
9.9	83	3	0.6	64.5	60	332	1.9	2.8	1	0	12.1
10.6	88	1.8	2	14.3	60	232	4.7	9.2	3	354	11.9
6.2	98	4.2	1.8	84.8	60	4	8.4	13	4	354	11.9
5.4	92	0.6	1.7	8.4	60	348	5.6	8.4	2	0	12.1

A	B	C	D	E	F	G	H	I	J	K	L
11.7	84	9.6	1	15.7	60	211	4.1	7.9	72	354	11.9
11.8	90	12.6	8.5	24.3	60	210	5.8	9.8	56	354	11.9
11.9	87	10.2	4	69.2	60	260	2.7	5.4	35	0	12
13.2	94	34.8	16	71.9	60	251	5.1	8.9	3	354	11.9
12.8	90	3	5.2	84.7	60	344	1.9	4.8	0	0	12.1
12.6	90	19.2	9.8	37	60	228	4.2	6.3	3	354	12
12.8	85	1.8	1.8	73.3	60	289	1.4	3.7	0	0	12.1
6	74	3	1.3	10.3	60	110	2.8	4.7	5	0	12.1
12.7	92	9	8.4	45.5	60	219	4.9	8.1	3	354	12
12.7	87	8.4	3.7	77	60	290	3.3	7.8	0	0	12
13.1	91	12	10.4	55.8	60	225	5.2	9.7	3	354	11.9
12.9	87	7.2	2.5	79.5	60	309	2	4.7	0	0	12.1
6.6	97	8.4	12.9	84.8	60	4	6.3	11.7	3	354	12
5.9	93	2.4	6.7	6.7	60	354	3.5	7	2	0	12.1
0.7	94	3.6	3.7	23.3	60	281	2.5	4.4	6	354	11.9
1.2	90	3.6	3.4	9.2	60	338	1.3	2.6	4	0	12
3	78	0.6	0.2	8.9	60	33	7	8.3	2	359	12.1
1.7	86	1.2	0.5	4.5	60	343	2.7	4.7	3	0	12.1
1.5	92	1.8	0.4	19.6	60	316	1.2	2.7	4	353	11.9
1.3	88	3	1.3	5.8	60	318	1.2	2.3	3	0	12
5.7	73	0.6	0.1	9	60	108	4.2	6	41	0	12.1
0.6	95	1.2	2	25.4	60	293	3.8	7.8	24	353	11.9
0.7	91	2.4	3	12.2	60	345	1.7	2.7	11	359	12.1
15.9	92	20.4	2.7	29.1	60	204	5.1	9.6	2	354	11.9
15.6	88	3.6	0.6	16.1	60	295	1.2	1.8	0	0	12.1
15.8	89	0.6	2.1	18.2	60	105	1.7	2.3	0	0	12.1
15.3	84	1.2	0.7	41.6	60	195	5	6.7	50	354	12
15.1	92	2.4	4.6	46.2	60	195	5.1	10.8	16	354	11.9
14.8	85	0.6	1.3	20.5	60	59	0.7	1.7	5	0	12
15.1	94	1.2	2.2	48.4	60	204	3.9	6.5	3	354	11.9
14.9	89	1.8	1.3	21.7	60	0	0.6	2.9	1	0	12
14.6	91	13.2	9.2	111.4	60	156	6.2	9.3	3	354	11.9
13.7	92	2.4	5.6	46.1	60	102	2.9	4.8	0	0	12.1
9.9	83	0.6	0.3	50.6	60	286	2.2	5.2	31	0	12.1
8.7	81	4.2	1.2	51.9	60	329	2.1	5.6	81	0	12
15.3	97	18	7.9	56.3	60	182	5.9	9.6	2	354	11.9
14.7	89	3.6	2.3	24.1	60	70	1.5	3.6	0	0	12
14.4	92	12.6	10.9	67.2	60	195	5.5	9.2	2	354	12
14.5	87	1.8	2.9	26.9	60	0	1.1	2.3	0	0	12.2
14.5	99	9.6	10.3	77.5	60	166	4	7.2	3	354	11.9
14.1	92	17.4	5	31.9	60	111	3.3	5.4	0	0	12.2
12.9	93	3	3.2	35.1	60	101	4.3	8.2	0	0	12
14.7	88	0.6	1	82.5	60	184	4.5	9.4	2	354	11.9

A	B	C	D	E	F	G	H	I	J	K	L
15.2	87	3.6	8	92.1	60	170	7.6	10.4	2	354	11.9
14.7	84	1.2	1.2	37.3	60	106	2.3	4.8	0	0	12
15.1	89	10.8	10.1	102.2	60	171	4.5	7.5	2	354	11.9
14.6	86	7.2	3.2	40.5	60	105	4.4	5.8	0	0	12.1
13	85	21.6	2.1	117.8	60	202	6	11.1	3	354	12
9.6	80	8.4	0.3	126	60	240	5.5	9.9	32	353	11.9
8.4	81	14.4	2.3	128.3	60	0	0	0	59	0	11.9
5.6	92	0.6	0	0	60	144	7.6	10.6	4	354	11.9
5.8	88	0.6	1.6	53.8	60	109	4.8	8.4	2	0	12
4.7	82	25.8	0.5	0.5	60	228	5.4	7.9	4	354	11.8
5.2	84	10.2	5.1	65.9	60	0	0	0	2	0	12.1
4.7	82	25.8	0.5	0.5	60	228	5.4	7.9	4	354	11.9
5	72	22.2	2.4	2.4	60	218	5.3	6.6	4	354	11.9
5.8	69	0.6	2.2	57.9	60	289	2.9	5.8	2	0	12
4.7	82	25.8	0.5	0.5	60	228	5.4	7.9	4	354	11.9
5.7	87	9	7.5	73.4	60	267	3.2	5.3	2	359	12.1
5.7	87	0.6	1.9	1.9	60	281	3.2	8	2	0	12
7.2	98	0.6	6.6	33	60	222	5	8.6	20	353	11.9
6.6	96	10.8	5.2	26.5	60	213	6.5	10.8	4	353	11.9
6.3	89	18.6	1.9	4.3	60	0	0	0	2	0	12
7.7	95	1.8	4.2	37.2	60	221	5.4	9.7	25	353	11.9
10.7	94	4.8	0.6	38.5	60	258	6	8.1	15	354	11.9
15	98	3	8.2	47.9	60	171	4.5	7.3	11	355	11.9
13.2	98	7.2	3	3.2	60	81	2.4	6.4	6	0	12.1
15.4	100	70.8	26.8	74.7	60	211	11.9	20	21	354	11.9
14.5	98	4.8	8.8	12	60	66	3	4.7	12	0	12
13.9	88	1.8	6.4	81.1	60	214	6.4	9.6	60	354	12
1.6	99	26.4	4.9	4.9	70	86	16.1	17.7	10	354	11.9
13.1	83	0.6	0.6	82.9	60	193	6.1	11.1	15	355	11.9
2.6	98	22.8	13.3	27.3	60	117	7.7	9.5	4	353	11.8
0	97	16.2	0.6	0.6	70	80	15.8	19.5	32	354	11.9
-0.1	84	0.6	2.5	2.5	60	0	0	0	21	0	12
0.6	99	25.2	4.2	4.2	60	75	15.2	18.3	21	354	11.9
4.1	95	2.4	1.4	1.4	60	9	7.6	13.4	4	354	11.9
1.7	99	3	2	2.1	70	88	11.5	14.7	13	354	11.9
0	83	9.6	1.6	1.6	70	85	13.8	14.6	11	354	11.9
5.4	88	4.2	2.2	85.8	60	134	5.3	6.7	13	354	11.9
5.1	83	1.8	0.3	13.5	60	107	2	4.3	9	0	12
1.6	99	26.4	4.9	4.9	70	86	16.1	17.7	6	354	11.9
0	83	9.6	1.6	1.6	70	85	13.8	14.6	5	354	11.9
6.2	100	0.6	0	91	60	13	5.5	7.8	4	354	11.9
4.3	100	0.6	2.8	9.7	60	356	8.1	12.8	4	353	12
5.7	100	21	5.6	97.1	60	1	4.4	7.5	4	354	12

A	B	C	D	E	F	G	H	I	J	K	L
5.8	97	1.2	0.2	17.1	60	350	3.5	5.3	2	0	12.2
4.8	100	13.8	4.1	6.9	60	3	6.9	10.8	4	353	11.9
0	97	16.2	0.6	0.6	70	80	15.8	19.5	17	354	11.9
0.4	83	2.4	0.8	21.3	60	94	10	12.5	16	0	12.1
0	97	16.2	0.6	0.6	70	80	15.8	19.5	28	354	11.9
0	89	4.2	4.9	7.4	60	87	10.1	12.2	18	0	12
1.8	95	17.4	4.6	4.6	60	92	9.1	11.3	3	0	12.1
3.1	98	7.8	2.1	2.1	60	130	6.5	8.2	4	354	11.9
2.9	94	1.2	1.2	1.2	70	108	3.4	6.2	3	0	12
1.4	99	9	3.6	3.7	70	0	0	0	21	0	11.9
1	95	7.8	1.4	1.4	60	85	9.5	10.9	14	0	12.1
1	95	7.8	1.4	1.4	60	85	9.5	10.9	11	0	12.1
-1.9	87	0.6	0.3	4.3	60	285	1.8	2.8	13	353	11.8
-1.9	89	1.8	0.9	4.6	60	339	0.7	1.5	8	0	12.1
-2.1	91	1.2	1.5	5.7	60	279	3.4	4.8	46	353	11.9
-1.9	89	1.8	1.1	5.7	60	277	0.9	2.1	24	0	12.1
2.1	62	2.4	0.1	6.6	60	111	3.5	4.5	72	0	12
4.9	92	3	1.4	8.4	60	160	2.4	3.8	48	354	11.9
5	89	1.2	0.6	7.4	60	107	3.5	6.5	41	0	12.2
7.1	98	0.6	0.8	10.3	60	199	2.1	3.1	4	354	11.9
6.7	89	0.6	0	9.3	60	108	2.3	4	2	0	12
7.1	98	4.2	1.9	12.2	60	188	1.6	5.1	4	354	11.7
7.3	92	1.2	4.3	13.6	60	112	2.9	4.2	2	0	12
1	99	3	2.7	25.8	60	1	4.7	8	4	353	11.9
1.2	94	3	2	27.9	60	352	2.9	4.3	3	0	12
2.6	97	1.8	0.5	13.6	70	21	2.3	3.8	35	354	11.9
2.4	96	0.6	1.8	15.4	60	12	4.2	6.2	29	353	11.9
1.8	92	0.6	0.4	14.9	60	5	3.1	4.3	24	0	12.1
1.3	99	2.4	1.3	30	60	351	5.8	9.3	4	354	11.9
1.4	97	1.8	1.8	17.3	70	8	3.8	6.1	15	354	11.9
0.8	98	3.6	3	20.3	60	3	3.3	6.3	7	353	11.9
0.8	99	2.4	2.8	23.1	60	8	3.9	6.7	4	354	11.9
1	93	3.6	3.2	18.1	70	356	2.4	4.4	13	0	12.1
1.3	99	0.6	2.8	28.6	60	357	5	8.8	5	353	11.9
1.4	92	3.6	2.4	30.2	60	4	3.9	7.1	3	359	12
-3.7	93	1.8	1.7	34.9	70	241	4.5	6.4	5	353	11.9
-3.8	91	0.6	1	37.6	60	281	2.5	7.3	4	0	12
-6.6	78	0.6	0.1	34.6	70	108	3.6	6.2	4	0	12
-5.5	90	1.2	0.5	31.8	70	194	5.1	7.1	5	353	11.8
-5.3	88	0.6	0.6	35.6	60	94	2.3	4.4	4	0	12
-4.6	92	1.2	1.4	33.2	60	211	3.7	5.6	5	353	11.9
-4.6	89	1.2	1	36.6	60	331	1.6	2.8	4	0	12
-3.6	91	0.6	0.7	35.6	60	245	4.2	8.2	5	354	11.9

A	B	C	D	E	F	G	H	I	J	K	L
-3.8	90	0.6	0.7	38.3	60	260	4.1	6.8	4	0	12.1
-4.2	89	0.6	0.4	38.8	60	271	2.3	5.3	4	0	12
-11.3	67	0.6	0.1	39.6	70	282	2.1	3.9	60	0	12
3.9	90	4.8	4.9	48.3	60	261	2.6	4.1	42	354	11.9
5	85	0.6	0	36.3	60	233	3.1	6.2	62	354	11.9
5.4	82	1.8	0.2	40.4	60	274	2.8	6.6	38	0	12.1
4.9	92	7.2	4.8	41.1	60	250	4.1	5.8	44	353	11.9
5.1	86	0.6	1.4	41.7	60	278	1.5	2.8	29	359	12.1
-9	84	0.6	0.3	49.3	70	241	1.6	2.3	6	354	11.9
4.5	94	4.8	1.9	51.7	60	204	3.1	6.4	4	354	11.9
5.6	77	2.4	0.1	53	60	215	3.2	4.7	4	354	11.9
7	64	0.6	0.1	44.6	60	301	3	6.5	1	0	12.1
9.2	89	1.2	0.7	45.5	60	310	1	5.5	22	0	12.1
9.4	94	1.2	1	54.2	60	254	4	5.9	34	354	11.9
9.2	88	0.6	0	44.8	60	287	1.3	3.6	26	0	12
9.6	94	1.2	1.2	56.6	60	270	4.1	5.9	32	354	11.8
2.3	87	22.2	9.5	9.5	60	120	8.7	10.7	41	354	11.9
2	97	15.6	5.4	5.5	60	63	9	10.6	15	354	12
1.9	91	7.8	5.8	10.6	70	93	3.4	4.2	7	0	12
2	89	24	2.5	2.5	60	108	9.1	11.9	15	354	11.8
1.9	98	42.6	5.4	5.4	60	77	9.8	10.2	4	354	11.9
2.5	98	25.2	29	34.3	60	98	8.5	10.3	4	354	11.9
-8.8	78	1.2	0.4	7.2	70	112	3.4	5.4	90	0	12.1
-8.4	84	0.6	1.4	8.6	70	120	2.2	4.8	185	0	12
-7.7	84	0.6	0.8	11.4	60	111	3.6	6.6	4	0	12.1
-7.7	83	0.6	0	11.1	60	168	3.4	4.7	5	353	11.9
-7.6	83	0.6	0.5	11.9	60	196	3.2	5.4	5	353	11.9
-7.2	80	0.6	0.4	12.2	60	96	2.3	4.9	4	0	12.1
-0.9	92	1.2	0.6	12.9	70	138	0.4	0.8	4	354	11.9
-1.5	91	2.4	1.2	14	60	101	1.2	1.7	3	0	12.1
-7.9	85	0.6	0.3	10.9	70	163	5.3	8.7	5	353	11.9
-8	84	1.2	0.6	10.5	70	114	2.2	5.2	4	0	11.9
-8.1	82	0.6	0.3	9.9	70	111	3	6.6	12	0	12
-7.6	84	0.6	0.1	11.4	70	175	2.8	4.3	5	353	11.9
-0.2	95	0.6	1.3	14.2	60	280	3.1	4.1	4	354	11.8
-1.5	78	0.6	0.1	12.5	70	45	0.8	1	30	0	12
0.8	99	3	0.5	0.5	60	6	9.9	19.1	58	353	11.8
0	95	1.8	0.9	15.1	60	23	9.9	16.5	57	353	11.9
0.4	98	1.8	0.5	0.5	60	280	999.9	999.9	56	354	11.9
0.4	93	0.6	0.5	16.2	60	6	9.1	13.5	51	0	11.8
0.6	98	0.6	0.7	0.7	70	2	9.2	17.3	78	353	11.9
0.9	99	1.2	0.4	3.2	60	355	10	15.7	4	353	11.9
0.8	88	0.6	0.6	18.7	60	348	5.7	13.4	3	0	12.1

A	B	C	D	E	F	G	H	I	J	K	L
0.5	79	0.6	0.1	17.3	70	33	11.8	14.2	23	354	11.9
-0.9	86	2.4	8.4	8.4	60	85	7.9	9.3	4	353	11.9
-0.9	88	1.2	2.2	25.5	70	87	6.6	8.9	3	0	12.1
7.7	74	31.8	3.6	8.2	60	289	5.9	11.9	3	354	12
4	72	1.2	0	19.5	60	7	7.5	10.2	0	0	12
12.3	63	0.6	0.1	4.5	60	248	6.8	9.5	9	355	11.9
-1.1	86	0.6	0	23.2	60	74	7.8	10.5	3	0	12.1
-2.9	89	0.6	0.6	28.2	60	3	8.8	12.7	48	0	12
-1.1	90	13.8	5.5	5.6	60	88	8.6	10.7	4	354	11.9
-1.3	92	1.2	0.6	30.3	60	111	2.5	4.9	3	0	12.1
-2.1	88	0.6	0	27.6	60	21	7.7	9.3	51	0	12
-4.2	82	0.6	0.3	28.8	60	1	5.9	8.3	58	0	12
-1.5	93	0.6	0.6	13.6	70	123	4.1	5.2	15	354	11.9
-1.5	91	0.6	0.2	30.7	60	112	1.7	3.8	9	0	12.1
-1.9	85	0.6	0.3	11.7	60	136	0.7	1	4	353	11.8
-1.8	90	1.2	0.5	29.7	70	109	2.3	3.5	3	0	12.1
-1.4	91	0.6	0.1	14.1	70	111	4.6	5.6	82	353	11.9
-2.1	92	0.6	1.7	16.1	60	191	4.8	6.7	4	353	11.9
-1.1	91	0.6	0.2	14.3	70	120	4.2	5	153	353	11.8
-1.2	71	0.6	0.1	32	70	0	1.4	5.4	3	0	12.1
14.7	78	1.8	0.4	18.3	60	211	3.9	6.9	2	354	11.9
14	84	0.6	0.7	34.5	60	0	0	1.2	0	1	12.1
5.4	100	0.6	4.3	35.5	60	77	3.7	4.6	63	354	11.9
5.3	94	10.8	2.4	42.1	60	149	1.8	3.1	58	0	12.1
5.2	100	6	1.3	38.6	60	60	3.3	4.9	62	354	11.9
5.2	94	1.2	1.5	46	60	190	0.3	1	51	0	12
5.1	93	0.6	0.4	20.5	60	50	1.9	2.5	4	354	11.8
6	97	1.2	0.5	21	60	78	2.1	2.7	3	354	11.9
5.4	100	0.6	2.2	31.2	60	78	5.8	6.7	94	353	11.9
5.2	95	0.6	1.6	39.7	60	122	1.2	2.9	57	0	12.1
6	99	0.6	0.7	21.7	60	74	4.9	5.6	4	354	11.9
5.8	99	1.8	0.9	22.7	60	71	5.1	5.9	4	354	12
6.3	94	1.2	0.4	35.9	60	143	1.4	2.6	2	0	12
5.6	99	6	1.5	24.1	60	27	4.7	6.3	13	354	11.9
5.7	93	0.6	0.6	36.5	60	152	0.8	1.3	15	0	12
5.4	99	1.2	1	26.1	60	63	5.6	6.7	49	354	12
5.3	94	0.6	0.7	38.1	60	139	1.2	2.5	32	0	12.1
5.4	100	0.6	1.8	37.3	60	66	3.5	4.6	79	354	11.9
5.4	95	1.2	2.4	44.5	60	89	0.9	2.2	45	0	12
5.5	100	0.6	0.4	39.8	60	9	2.2	3.7	4	354	11.9
5.7	100	1.8	1.2	41.6	60	5	2.8	5.2	4	354	11.9
5.2	95	1.2	0.8	46.8	60	127	0.9	2	36	0	12.1
5.4	100	0.6	0.7	40.4	60	4	2	3.9	4	354	11.9

A	B	C	D	E	F	G	H	I	J	K	L
5.4	97	0.6	0.4	47.6	60	6	1.1	1.5	2	0	12
7	100	0.6	0.3	42.3	60	330	0.7	1	3	354	11.9
6.3	95	0.6	0.4	48.4	60	15	0.9	1.2	2	0	12.1
12	92	94.8	7.3	8.1	60	160	5.7	9	3	354	11.9
11.1	90	5.4	2.6	51.6	60	102	3.9	6.7	1	0	12
9.4	84	1.2	0.5	15.3	60	259	6	9	3	354	11.9
8.6	86	35.4	5.4	5.4	60	262	8.9	12.2	3	354	11.9
6.6	44	0.6	0.1	55.8	60	320	3.2	5.9	0	0	12.1
3.6	99	5.4	0.3	0.3	60	5	8	15.9	40	354	11.9
2	91	13.2	7	10.9	60	334	4	5.9	42	0	12.1
5.3	99	1.2	3.4	11.3	60	19	3.5	6.6	33	353	11.9
5.1	94	3	1.4	60.4	60	27	1.2	2	25	0	12
5.9	98	2.4	2.4	7.8	60	89	1.4	1.7	16	354	11.9
4.9	94	1.8	1.3	59	60	39	2.6	3.4	13	0	12.1
4.5	99	14.4	7.3	18.6	60	15	5.6	9.1	40	353	11.9
3.5	95	8.4	4.9	67	60	2	5.2	8.1	26	0	12.1
3.6	99	5.4	0.3	0.3	60	5	8	15.9	64	354	11.9
4.9	100	4.2	0.6	4.4	60	17	2.3	4.3	4	353	11.9
4.5	96	1.2	0.8	56.7	60	33	1.7	2.5	2	0	12.1
5.4	99	0.6	0.3	5.5	60	15	2.4	4.6	7	354	11.9
4.9	95	3	0.8	57.7	60	354	1.7	5.2	3	0	12.1
4.5	99	14.4	7.3	18.6	60	15	5.6	9.1	28	353	11.9
1.8	91	0.6	2.2	4.7	60	345	5.3	8.6	24	354	11.9
2.4	90	2.4	0.4	2.5	60	356	6	9.7	34	353	11.9
1.7	89	3	0.8	3.6	60	5	3.3	6.8	22	0	12.1
8	90	3	6	6	60	309	2.6	4.7	12	354	11.9
8	87	1.8	1.4	7	60	321	2	4.5	6	0	12.1
7.2	89	3.6	3.1	3.1	60	312	3.4	6.8	4	354	11.9
7.6	87	3.6	1.9	8.9	60	348	2.4	4.3	2	0	12.2
7.2	90	1.2	5.9	9	60	337	2.8	4.8	3	354	11.9
6.8	84	0.6	0.7	9.6	60	352	2.8	5.7	1	0	12.1
6	47	0.6	0	9.6	60	121	1.3	3.8	109	0	12
13.3	84	1.8	0.6	12.1	60	257	2.9	6.9	0	0	12.1
13.7	92	28.2	7.2	24.6	60	207	6	9.4	3	354	11.9
13.9	88	2.4	5.7	17.8	60	269	2.8	6.2	0	0	12.1
14.8	92	1.2	2.4	28.5	60	222	6.1	11.7	2	354	11.9
5	91	0.6	0.7	25.3	60	0	3.3	5	2	0	12
3.5	90	0.6	0.2	41.4	60	30	6.3	8.5	4	354	11.9
3.6	77	0.6	0.1	26.2	60	79	2.2	2.7	239	0	12.1
6.1	81	11.4	3.9	46.1	60	288	6.3	10.9	18	353	11.9
-0.1	92	1.2	0.2	0.2	60	331	9.5	16.4	136	353	11.9
1.8	64	1.2	0	7.3	60	281	6.8	12.5	256	354	11.9
-0.6	77	5.4	0.3	0.5	70	325	9.3	16	30	354	12

A	B	C	D	E	F	G	H	I	J	K	L
-1	73	3	0.1	0.6	60	332	7.8	15.2	62	353	11.9
-0.2	62	0.6	0	6.6	70	173	5.2	10.5	4	353	11.9
3.7	51	1.2	0	30.1	60	103	2.3	5.5	3	0	12.1
7	71	1.2	1	34.6	60	262	3	5.8	42	0	12.1
6.5	89	7.2	0.7	3.3	60	211	7.8	10.4	58	354	11.9
7.1	93	26.4	3.3	3.3	60	207	4.8	8.2	62	354	12
2.9	78	1.2	0	5.4	60	349	2.7	4.4	2	0	12.1
7.5	74	3	1.3	35.9	60	285	3.4	6.8	49	0	12.1
6.4	60	0.6	0.5	33.6	60	292	1.6	4.2	47	0	12.1
7.7	84	4.2	4.1	40	60	281	1.8	6.1	45	1	12.1
8.5	88	1.2	1	1.4	60	283	1.7	4	74	0	12
9.2	88	1.2	0.9	2.3	60	264	2.1	3.9	56	0	12
8.5	90	1.2	1.3	4.4	60	348	1.2	1.9	3	0	12.1
1.9	74	0.6	0.2	9.9	60	335	2.5	5.9	58	0	12.1
-1.8	77	1.8	0.1	10.2	60	7	7.7	12.7	3	0	12
-0.5	93	3	1.4	11.5	70	15	5.3	13.6	3	0	12.1
0.4	63	0.6	0.8	0.8	70	180	5	10.2	21	354	11.9
-0.2	62	0.6	0	6.6	70	173	5.2	10.5	13	353	11.9
14.6	87	2.4	5.3	6.6	60	259	2.3	3.7	2	354	11.9
14.4	82	1.2	2.5	6.2	60	308	1	3.4	0	0	12
14.3	90	0.6	4.6	12.4	60	192	3	6.4	2	354	11.9
15	63	1.2	0.2	1.3	60	214	2.9	5	1	354	12
12.9	92	0.6	1	12.7	60	117	2.8	4.6	0	0	12.1
13	62	0.6	0.1	0.9	60	155	3	4.8	56	354	11.9
14.4	91	1.2	1.3	7.9	60	208	1.9	5.2	2	354	11.9
14.5	81	1.8	1.2	7.4	60	255	1.9	4.6	0	0	12
14.2	85	1.2	1.3	8.6	60	91	2.7	3.9	0	0	12.1
14.4	93	2.4	3.1	15.5	60	179	2.5	5	2	354	11.9
13.5	88	0.6	2.5	11.2	60	105	2.5	4	0	0	12.2
7.8	82	0.6	2.6	2.6	60	80	8.7	10.5	4	354	11.9
6	89	8.4	5.8	8.4	60	24	6.4	10	13	354	12
6.9	85	7.2	3.4	8.5	60	112	3.2	6.8	2	0	12
7.4	82	12.6	1.6	5	60	108	3.5	5.8	2	0	12.1
6.8	92	4.8	0.2	0.2	60	19	5.8	8.5	4	353	11.9
7.3	88	4.2	2.1	2.1	60	66	6.7	8.4	4	353	11.9
6.7	93	0.6	1.2	17.1	60	65	4.6	5.3	2	0	12
7	95	2.4	4.1	6.7	60	339	1.8	2.4	4	354	12
6.4	95	1.8	3.2	20.3	60	44	2.1	2.8	2	0	12
7.4	86	1.2	0.3	7.2	60	89	3.8	6.8	112	353	11.9
7.2	95	1.2	4.6	4.6	60	86	10.1	13.4	209	354	11.9
6.9	92	0.6	3.3	29.8	60	95	6	8.6	89	0	12
7	86	1.8	3.2	10.4	60	105	4.9	8.4	85	353	11.9
6.7	84	4.2	1.6	22.2	60	108	3.2	5.2	89	0	12

A	B	C	D	E	F	G	H	I	J	K	L
6.6	88	7.2	2.2	2.2	60	85	6.9	9.4	111	354	11.9
6.5	87	1.8	2.4	24.6	60	99	3.3	5.2	84	1	12.1
6.6	88	7.2	2.2	2.2	60	85	6.9	9.4	73	354	11.9
6.3	89	4.2	1.8	26.4	60	94	3.4	4.8	57	0	12.1
7.5	96	3	7.2	11.8	60	94	6	10.3	35	354	11.9
7.1	92	1.2	0.6	30.4	60	94	4.6	6.6	34	0	12.1
7.6	99	5.4	12.9	12.9	60	80	8.7	10.6	45	354	12
7	94	3	5.2	35.6	60	83	5	6.7	24	0	12.1
7	95	0.6	2.3	37.9	60	70	5.4	8.3	14	0	12.1
7	96	4.2	2.4	2.4	60	74	5.1	7.5	6	0	12.1
7.2	100	0.6	4.9	4.9	60	64	8.6	10.1	8	354	11.9
7.1	96	3.6	0.8	5.2	60	76	5.1	7.7	2	0	12.1
8	100	3.6	9.6	14.3	60	62	5.8	7.9	4	354	12
6.9	95	7.2	3	11.8	60	354	3.5	7.2	4	354	12
6.8	93	2.4	0.1	5.7	60	15	5.6	8.1	2	0	12.1
10.7	71	7.2	0.4	14.7	60	247	7.2	12.9	2	354	11.9
10.8	81	4.2	2.2	15.2	60	344	4.2	6.2	0	0	12.2
8	95	0.6	3	33	60	5	5.1	8.9	3	353	11.9
10.4	94	4.2	11.6	26.3	60	310	1.6	3.4	3	354	11.9
9.7	89	1.8	4.7	19.9	60	9	2.9	4.2	1	0	12.1
9	96	9	3.7	30	60	7	4.3	7.5	3	354	11.8
8.6	93	0.6	2	21.9	60	21	3.4	4.4	1	0	12.1
15.6	67	11.4	1	23.2	60	339	2.3	4.9	9	0	12.2
18	86	6.6	3.6	3.6	60	309	0.9	2.8	0	0	12.2
19.1	90	7.2	5.7	9.3	60	340	2.1	4.2	0	0	12
16.9	96	1.2	1.3	25.2	60	19	1.4	2.9	1	354	11.9
18.4	89	16.2	2.8	16.1	60	289	0.9	1.7	0	0	12
20.2	88	5.4	1.5	24.9	60	274	1.4	6.8	-2	0	12.1
20.5	88	3	1.2	26.1	60	233	1	2.5	-2	1	12.1
17.7	92	5.4	1.6	111.1	60	294	0.7	0.9	8	354	11.9
17.4	89	3.6	1.2	41.3	60	0	0	0.5	5	0	12.1
18.7	94	1.8	29.4	88.1	60	343	4.9	7.6	14	354	11.9
17.6	76	13.2	0.7	27.1	60	307	1.1	3.7	3	1	12.1
17.9	93	8.4	4.1	92.2	60	307	1.7	4.5	11	354	11.9
18	87	1.2	4.8	31.9	60	344	2.1	3.5	11	0	12.1
17.2	96	6	11.6	103.8	60	22	3	5.5	22	354	11.9
16.8	93	16.8	5	37	60	92	0.9	1.9	15	0	12.1
16.4	99	3	4.4	115.5	60	157	1	3.6	11	354	11.9
16.9	93	1.2	2.3	43.6	60	0	0	0.5	12	0	12.1
15.8	99	0.6	2.5	118	60	131	0.3	0.8	104	354	11.9
16.7	95	0.6	3.9	47.5	60	119	1.2	2.6	94	0	12.1
18.2	84	0.6	0.1	48	60	71	1	2.4	40	1	12.1
17.6	74	2.4	1.1	128.8	60	261	3.3	5.2	152	355	11.9

A	B	C	D	E	F	G	H	I	J	K	L
15.5	85	9.6	4.5	124.2	60	9	5.8	10.8	1	354	12
17.7	81	0.6	0.1	129.9	60	325	0.9	1.7	68	354	11.9
17.2	81	1.8	0.6	50.3	60	342	1.4	3.3	30	1	12.2
17.4	84	0.6	3.6	53.9	60	335	1.2	2.2	158	0	12
17.1	96	0.6	3.5	134.4	60	67	2.1	3.9	2	354	11.9
17.1	95	0.6	0.8	135.5	60	86	3.1	6.5	2	354	12
16.3	89	1.2	2.1	58.2	60	139	1.3	3.3	0	0	12.1
22.6	98	15.6	13.9	712	60	219	1.6	3.5	5	354	11.9
22.3	94	8.4	5.8	250.1	60	50	0.7	1.6	5	1	12.1
20.4	98	3	13.6	151.3	60	203	2	3.6	1	354	12
20.4	91	8.4	4.6	65.9	60	224	0.8	1.5	-2	0	12
20.6	97	0.6	2.7	154.1	60	223	2.7	5.8	1	354	11.9
21	91	5.4	1.1	137.7	60	264	4.4	6.7	0	354	12
20.9	90	4.8	2.3	61.3	60	274	2.2	6	-2	1	12.1
22	97	4.8	13.6	725.6	60	223	3.5	8.1	86	353	11.9
19.5	78	1.8	0.1	162.2	60	283	2.4	5.8	45	354	12
19.2	71	1.2	0	69.9	60	348	1.5	3.3	43	0	12.1
19.9	95	1.2	7	169.2	60	177	3	5.7	123	354	11.9
20.3	96	33	7.7	186.5	60	35	12.9	19.5	1	354	11.9
19.7	96	5.4	5.7	79.3	60	68	2.5	6.1	0	0	12.1
22.9	99	2.4	28.2	218.7	60	258	2.6	5.2	18	354	12
19.9	94	4.8	1.8	247.3	60	259	3.2	5.3	1	354	11.9
20.1	88	7.2	1.6	92.5	60	264	2	3.9	-2	0	12.2
20.2	94	7.2	15.4	285.1	60	259	7.6	11.5	34	354	11.9
20.5	94	0.6	5.7	290.8	60	243	4.6	7.4	92	355	11.9
20.3	97	17.4	11.1	258.4	60	244	2.7	3.7	2	354	12
20.3	97	22.2	11.3	269.7	60	256	5	6.7	6	354	11.9
19.7	87	15	9.5	102	60	256	1.8	3.5	-2	0	12.1
22.5	88	10.2	4.1	121.7	60	72	1.8	2.6	7	1	12.1
23	93	4.2	1.4	292.3	60	23	1	1.3	91	354	12
22.7	77	0.6	0.1	125	60	126	1.2	2.4	32	1	12.1
23.7	71	3	0.1	127.6	60	259	2.2	4.1	-3	1	12.1
24.7	93	0.6	7.1	301.4	60	219	2.1	3.5	120	355	12
20	96	3.6	31.8	333.2	60	158	1.2	3.4	2	354	12
19.4	91	1.2	9.4	346.7	60	194	6	9.6	87	354	11.9
21.5	88	5.4	16.4	363.2	60	233	5.2	9.3	1	354	12
22	96	1.8	15.8	379.1	60	221	2.8	5.2	0	354	12
21.9	96	0.6	6.6	385.7	60	236	2.2	3.8	0	354	11.9
22.1	91	0.6	1.5	137.9	60	0	0.5	2.1	-3	1	12.1
22.2	99	0.6	26.3	414.6	60	3	4.4	8.4	0	354	11.9
22.3	97	6	2.1	416.7	60	358	1.4	2.5	0	355	12
22.4	90	10.8	6.5	150.9	60	174	0.7	1.4	-3	1	12
21.1	93	3.6	4.9	421.6	60	112	1.2	2.1	1	354	11.9

A	B	C	D	E	F	G	H	I	J	K	L
22.2	87	3	6	156.9	60	162	1.5	3.9	-3	1	12.1
21.9	83	0.6	0.3	157.2	60	115	2	5.1	-3	1	12.1
25.6	83	12.6	0.7	158.1	60	118	1.6	3.2	-2	1	12.1
21.4	92	4.8	1.8	159.9	60	347	3.7	8.6	-2	1	12.1
21.8	100	1.2	17.2	441.7	60	46	3.1	6.5	1	355	11.9
22.1	78	10.2	0.3	164.2	60	257	2.4	3.9	-3	1	12
21.5	100	2.4	16.6	458.3	60	6	1.4	2.4	0	354	11.9
20.7	94	5.4	8.4	172.5	60	55	1.1	2	-2	1	12
22	100	2.4	3.3	461.7	60	243	2.5	4.3	0	354	12
21.7	95	13.8	3.9	176.4	60	0	0.3	1	-2	1	12
22	98	3	2.2	473.5	60	165	3.1	5.5	1	354	11.9
19.9	61	61.2	1.2	475	60	335	5.8	15.6	0	355	11.9
20.5	89	11.4	2.4	192.2	60	231	1.1	1.4	1	1	12.1
19.1	81	4.2	35.6	510.6	60	229	3.4	4.8	10	354	11.9
27.4	63	9.6	0.2	189.3	60	1	1.3	2.6	140	2	11.9
21.7	95	0.6	11.1	522.2	60	200	1.6	2.6	35	355	12
22.2	90	11.4	1	198.6	60	40	3.1	4.3	61	1	12.2
22.3	87	0.6	0	195	60	7	2.1	3.7	8	1	12.1
20.8	94	8.4	0.2	529.1	60	72	7.2	9.9	1	355	11.9
24.8	84	0.6	0	201.4	60	118	0.6	1.7	36	1	12
23.3	100	1.2	22.5	559.1	60	167	2.3	4.2	12	355	11.8
23.1	91	0.6	5.4	207.2	60	122	1	1.8	6	1	12
22	97	21.6	15.9	652	60	82	5.1	8.8	1	354	11.9
21.8	94	1.2	3.1	229.1	60	106	2.1	4	-3	1	12.1
23.8	98	9	2	600.7	60	203	2	3.4	0	354	12
24.3	97	3	1.9	572.2	60	0	0.4	0.7	0	354	11.9
24	88	0.6	1.4	209.8	60	340	0.3	0.6	-3	1	12
23.8	100	7.8	13.7	585.9	60	177	1	2	0	355	11.9
24	93	7.8	3.6	213.4	60	309	1.6	4.5	-3	1	12
23.9	98	5.4	7.9	593.9	60	260	1.9	3.5	0	354	11.9
23.9	92	5.4	7.3	220.6	60	251	0.9	2.8	-3	1	12
23.6	99	4.2	4.6	598.4	60	265	0.9	1.4	0	354	11.9
24	100	0.6	1.1	604.8	60	239	2.4	4.1	0	354	11.9
23.1	85	3	0.1	615.6	60	274	0.9	1.9	0	354	12
21.9	96	16.8	4.1	621.2	60	101	3.2	8	5	354	11.9
22.3	95	0.6	1.4	617.1	60	145	2.8	5.2	65	354	12
22	98	14.4	14.9	636.1	60	91	4.5	7.8	4	354	11.9
21.6	92	1.8	0.6	226.1	60	111	2.4	5.2	2	1	12
22.4	100	4.8	0.6	654.5	60	8	1.9	3.7	0	354	11.9
22.1	97	0.6	0.9	230.9	60	17	4	5.5	-3	1	11.9
22.9	100	0.6	1.3	655.8	60	5	3	5.6	0	354	11.9
23	93	183.6	4.6	662.2	60	249	3.7	5	1	354	12
21.8	89	8.4	1.5	234.9	60	313	0.9	1.7	-2	2	12

A	B	C	D	E	F	G	H	I	J	K	L
21.7	68	0.6	0	232.6	60	321	0.6	1.8	177	1	12
20.8	99	1.2	23.7	686	60	352	0.8	1.4	6	354	11.9
21.6	81	14.4	1.5	240.8	60	268	0.6	2.8	-3	1	12.1
20.4	74	0.6	0	241.6	60	100	0.6	2.1	-3	1	12.1
21.7	95	3.6	4.5	736.6	60	219	2.2	3.8	19	354	12
21.5	89	15	4.3	259.2	60	0	0.7	1.9	20	2	12
22.8	96	34.2	3.8	742.4	60	236	3.6	5.5	23	354	11.9
22.3	90	3.6	1.4	262.3	60	0	0.8	2	33	1	12.1
22.8	95	0.6	4.2	757.5	60	217	1.1	2.1	0	354	12
22.2	94	0.6	0.4	759.3	60	243	3.2	5.4	0	354	11.9
20.6	93	2.4	0.9	761.4	60	153	0.9	1.9	1	354	11.9
20	79	12.6	0.7	272.5	60	92	1	2.7	-3	1	12
20.6	97	2.4	4.1	765.5	60	121	0	0.8	1	354	12
20.3	90	5.4	3.3	275.8	60	115	0.9	2.4	-3	1	12.1
21.2	99	10.2	4.2	769.8	60	180	2.7	4	1	354	11.9
20.9	93	0.6	4.1	279.8	60	243	1.1	3	-2	1	12
21.1	99	1.8	3.1	772.8	60	170	0.8	1.4	3	354	11.9
20.9	95	1.8	5.1	284.9	60	112	0.6	2.3	1	1	12.1
21.1	96	3	8.9	293.9	60	100	1.1	2.5	2	0	12.1
21.7	100	0.6	1.7	778.4	60	164	2.5	6.1	69	355	12
21.5	96	0.6	2.8	296.7	60	67	0.6	1.3	23	1	12.1
23.7	99	2.4	2.5	783.8	60	0	0	0.5	13	354	12
22.1	94	0.6	5	302.5	60	43	0.7	1.3	18	1	12
23.4	100	0.6	1.7	785.5	60	107	0.8	1.3	71	354	11.9
22.1	95	4.8	7.2	818.5	60	269	4.9	7.1	0	354	12
21.9	91	0.6	3.4	315.1	60	261	3	5.4	-3	1	12
19.4	71	1.2	1.7	823.7	60	320	0.3	0.6	19	355	12
23.3	95	1.8	0.4	806.2	60	255	1.9	2.8	0	355	12
22.7	86	1.8	0.2	309.6	60	267	1.9	5.6	-3	1	11.9
23.3	89	3	4.2	794.7	60	264	5.8	10.2	0	354	12
24.7	56	4.8	0.4	306	60	252	2	3.9	24	2	12.2
23	96	3	4.7	811.4	60	251	4.3	7.1	0	354	11.9
22.5	90	2.4	2	311.7	60	246	1.5	4.2	-3	1	12
23.2	83	3.6	1.2	790.5	60	258	5.2	7.3	0	354	11.9
21.8	97	0.6	1.6	820.1	60	211	2	4.2	16	354	11.9
21	83	0.6	0.2	316	60	0	0.3	1.1	-3	1	12
18.2	88	0.6	0	317	60	44	3.6	4.4	-2	1	12.1
21.5	93	0.6	1	827.7	60	261	1.7	2.5	0	355	11.8
20.9	83	0.6	1.1	318.1	60	269	1.7	5.4	-3	1	12
21.8	86	3	0.3	826.7	60	248	4.8	7.9	0	355	12
19.3	93	0.6	0	828.4	60	250	2.4	3.7	1	354	11.8
19.3	90	20.4	11.2	839.9	60	36	4.4	7.2	10	354	11.9
18.7	92	9	4.2	324.9	60	185	0.7	2.4	7	1	12.1

A	B	C	D	E	F	G	H	I	J	K	L
15.8	89	1.8	0.8	345.7	60	71	5.2	7.5	26	1	11.9
20.2	69	1.2	0.6	328.1	60	252	3.1	7.4	38	1	11.9
18.3	87	7.2	1.5	320.7	60	16	2	4.7	1	1	12
10.9	70	0.6	0	857.5	60	333	1.4	3.1	2	354	12
16	87	3	2.7	902.3	60	38	9.3	14.2	14	354	11.9
15.1	74	4.2	0.8	329.2	60	254	0.8	3.4	0	1	12.1
11	88	2.4	1.7	333	60	355	3.4	5.6	111	1	12
16	86	7.8	1.2	899.6	60	47	6.4	10.7	15	354	11.9
11.8	87	1.2	11.7	874.8	60	344	6.3	9.6	49	354	11.9
15.7	84	0.6	0.1	343.9	60	65	6.2	8.6	17	1	12.1
16	86	0.6	0.7	898.5	60	41	4	8.1	3	354	12
15.5	87	0.6	0	346.2	60	73	5.3	6.7	-2	1	12
15.3	93	0.6	0.8	918.8	60	320	0.9	2.4	3	354	11.9
14.5	98	9.6	19.3	943.1	60	59	1.9	2.9	5	355	11.9
15.5	94	1.2	5.5	352	60	69	3.1	4.8	9	1	12.1
16.1	97	1.2	5.9	949	60	85	2.1	3.8	86	354	11.9
15.5	94	1.2	0.8	352.9	60	73	1.7	2.5	50	1	12.1
14.7	94	6	5	923.8	60	60	4.9	9.2	8	354	11.9
15.4	90	0.6	0.1	346.5	60	345	1.3	2.4	7	1	12
18.7	98	12	5.6	978.7	60	168	1.9	3.1	41	353	11.9
18	92	2.4	5	372.1	60	113	2.3	3.5	27	1	12
18.4	93	0.6	0.2	372.3	60	112	2.3	4.8	133	1	12.2
17.4	92	1.8	0.5	950.5	60	221	1.3	2.4	1	354	12
16.9	80	5.4	0.7	354.7	60	57	0.4	0.7	-2	1	12.1
18.1	96	4.2	5	955.6	60	255	2	2.7	6	354	11.9
18.6	86	0.6	2.1	356.8	60	0	0	0.6	0	1	12.1
18.4	92	2.4	2.1	969.4	60	179	2.7	4.4	31	354	11.9
18.2	89	6.6	1.6	365.2	60	117	1.5	3	23	1	12
18.1	98	3	11.7	967.3	60	178	1.8	2.3	18	354	11.9
19.1	89	1.8	6.8	363.6	60	293	1.3	1.8	7	1	12.1
18.5	96	8.4	3.7	973.1	60	170	2.3	3.6	15	354	11.9
18.4	86	1.8	1.9	367.1	60	113	1.9	3.2	24	1	12.1
21.9	91	1.2	0.6	982.4	60	205	2.1	4	29	356	11.9
17.6	77	0.6	0.1	994.7	60	230	4.3	7.2	93	355	12
16.5	92	22.2	7	1009.5	60	278	4.7	7.7	13	355	11.9
16.2	90	0.6	4.1	386.6	60	0	1	2.9	4	1	12
13.4	68	1.2	0.2	993.6	60	194	2.1	3.2	2	355	11.8
17.7	85	5.4	1.9	996.6	60	246	4.5	7.5	43	355	11.9
17.3	91	1.2	3.4	1002.1	60	250	4.2	7.1	89	355	11.9
17.6	86	15	2.2	998.8	60	244	4	7.3	14	355	11.9
17.3	87	13.8	2.1	380.3	60	269	1.4	4.4	8	1	12.1
14.2	96	0.6	6.7	1016.2	60	306	1.2	4.2	12	355	11.9
13.5	91	10.8	3	389.6	60	345	2.1	3.5	2	1	12

A	B	C	D	E	F	G	H	I	J	K	L
18.4	83	24	4.6	1025.6	60	218	5.5	9.9	1	355	11.9
18.1	91	48	8.9	1034.6	60	249	4.1	7.7	1	355	11.9
18	90	9	4.6	396.8	60	304	1.7	9	-2	1	12.1
18.1	97	0.6	15.7	1050.3	60	244	2.5	3.8	1	355	11.9
14.7	79	0.6	0	1052.7	60	78	2.6	3.6	2	354	11.9
13	66	0.6	0.5	400.7	60	13	2.6	4.3	0	1	12.1
14.3	81	0.6	0.5	1053.2	60	112	1.1	2.1	2	354	11.9
7.5	84	0.6	0.6	401.3	60	21	3	5	1	1	12.1
7.9	86	1.2	1.6	1056.1	60	346	3.6	6.3	18	354	11.9
7.8	73	4.8	2.4	3.1	60	144	3.7	7	3	354	12
7.4	69	10.2	3.3	406.3	60	145	1.2	3	0	1	12.1
7.9	88	0.6	2.3	4.2	60	166	3.6	5.7	74	353	11.9
7.6	75	1.2	1	409.8	60	0	1.3	4.2	0	1	12
8	66	2.4	0.3	0.7	60	168	2.7	5.5	3	354	11.9
8	61	1.8	0.5	403	60	114	2.2	3.8	0	1	11.9
7.5	79	0.6	1.6	411.4	60	172	1.8	4.1	0	1	12
9.7	69	0.6	0	402.4	60	102	4.3	5.8	0	1	12
7.3	84	12	1	1	60	161	8.1	13.4	61	354	11.9
7.7	82	4.8	3.7	3.7	60	97	3.4	7	41	1	12.1
7.7	77	3	1.7	4.8	60	147	4	7.5	3	354	11.7
7.6	73	1.8	2.4	408.8	60	42	1.3	3	0	1	12.1
7.4	82	3.6	2.4	2.4	60	157	3.5	8.6	27	354	12
8.5	92	3	0.7	5	60	217	0.7	1.2	12	354	11.8
9	88	8.4	4.1	8.3	60	87	2.9	4.2	6	1	11.9
9.1	93	1.8	4.2	9.2	60	114	1.7	2.8	6	353	11.9
9.5	89	0.6	0.1	13	60	123	1.5	5.2	1	1	12.1
12.6	87	3	0.2	11	60	37	6.5	11.5	3	354	11.9
11.9	98	4.8	5.6	16.6	60	16	6.5	9.8	3	354	12
11.6	87	5.4	2.6	15.8	60	29	4.5	5.8	0	0	12.1
15.2	98	0.6	0.1	21.9	60	202	2.2	4.4	60	354	11.9
14.4	93	0.6	0.5	17.9	60	161	1.6	2.7	29	1	12
14.6	93	5.4	0.3	19.6	60	14	1.2	2.1	5	1	12
13.9	99	4.2	3.1	25.1	60	13	4.9	7.3	2	354	11.9
13.8	94	7.2	2	22.1	60	58	0.7	1.3	0	1	12.1
14	100	0.6	4.4	29.5	60	359	1.2	2.5	2	354	11.9
14.3	100	1.2	3.7	33.2	60	19	4.4	6.4	2	354	11.9
15.4	70	0.6	0.6	28.7	60	163	0.9	2.1	44	1	12.1
16.7	86	0.6	0.5	47.7	60	216	5.2	9.5	2	354	11.9
5.4	94	0.6	0.8	0.9	60	178	2.9	4.8	29	354	11.9
6.1	93	0.6	1.9	6.7	60	190	3.8	6	41	354	11.9
3.1	85	1.8	1	52.3	60	166	3.8	6	4	354	11.9
2.1	60	0.6	0	50.2	60	166	5	7.7	4	354	11.9
3.7	97	1.2	0.7	55	60	164	6.2	10.3	4	353	11.9

A	B	C	D	E	F	G	H	I	J	K	L
6	94	3.6	3.8	4.8	60	178	3.9	5.3	43	354	11.9
2.8	69	3	1.5	33.9	60	110	2.8	5	1	0	12.1
5	94	0.6	0.1	0.1	60	163	1.9	4.5	30	353	11.9
5.7	88	0.6	0.1	37.2	60	107	3.7	5.6	29	1	12
7.1	90	0.6	0.1	38.8	60	267	1.3	3.7	1	1	11.8
5.1	76	7.2	2	44.2	60	0	0.9	2.6	1	1	12.1
8.5	89	0.6	0.2	52.4	60	134	1.6	4.3	8	1	12.1
4.8	90	27	4.3	4.3	60	213	3.2	6.7	4	354	11.9
5.2	85	7.8	6.6	50.8	60	233	1.3	1.8	1	1	12.1
9.2	93	24	9.3	20.5	60	192	6.5	11	18	354	11.9
9.1	86	4.2	1.9	54.3	60	18	1.4	4	9	1	12.1
9.9	92	8.4	9	29.5	60	175	6.1	11.5	9	354	11.9
9.7	88	0.6	2.6	56.9	60	89	2.6	3.9	6	0	12.1
10	93	8.4	3.9	3.9	60	170	5.1	6.9	5	353	12
9.9	90	2.4	1.9	58.8	60	99	4.8	7.1	2	1	12.1
9.9	94	5.4	3.3	3.3	60	186	3.7	7.9	3	352	11.8
9.7	90	1.2	1.7	60.5	60	114	2.5	5.8	0	1	12.1
10.4	92	3.6	0.8	0.8	60	180	4.6	7.6	3	354	11.9
10.2	91	3.6	1.5	2.3	60	187	5	9.3	3	354	11.9
10.5	87	0.6	0.5	61.5	60	32	1.8	3.5	0	1	11.9
10.7	93	6.6	5.9	8.2	60	162	4.3	6.6	3	354	11.9
10.6	88	1.2	1.4	62.9	60	64	1.9	3.7	0	1	12.1
10.8	95	3.6	2.4	10.6	60	208	4.4	9.3	3	354	11.9
10.6	89	6	1.2	64.1	60	19	1.6	3.3	0	1	12.1
10.5	92	3.6	0.2	0.2	60	230	6.1	10.7	3	354	11.9
10.6	88	1.2	4.4	68.5	60	278	2.9	5.1	0	1	12.1
2.2	86	1.2	0.1	0.4	60	319	3.1	6.1	4	354	11.9
0.6	96	1.8	1.8	11.2	70	343	1.4	2.1	4	353	11.9
1.1	91	0.6	3.1	92.1	60	353	1.1	1.6	2	0	12
0.8	79	1.2	0.1	0.7	70	153	1.6	2.6	49	354	11.9
1.2	78	0.6	0.5	71.2	60	102	2.5	4.5	36	0	12
0.5	93	1.8	1.5	2.2	70	209	0.5	1.4	29	353	11.8
1	88	2.4	3.2	74.4	60	112	2.9	3.8	30	0	12
0.8	95	2.4	1.7	3.9	60	163	2.1	3.4	15	354	11.9
0.9	90	3.6	3.3	77.6	60	107	2.8	3.8	21	0	11.8
0.7	96	3.6	3.2	7.2	60	142	0.8	1.9	17	354	11.8
0.8	92	3	3.6	81.2	60	103	2.4	3.3	13	0	12
1.3	97	4.2	2.9	2.9	60	132	3.1	4.1	9	354	11.9
0.9	92	2.4	2.3	83.5	60	103	2.7	3.8	6	0	12.1
1.2	96	3.6	3.3	6.2	60	137	2.3	3.4	6	354	11.8
1	92	0.6	2.3	85.8	60	107	2.5	3.4	3	0	12
0.5	97	2.4	3.2	9.4	60	290	1.7	2.9	4	353	11.8
0.8	92	4.2	3.2	89	60	0	0	0.8	2	0	12

A	B	C	D	E	F	G	H	I	J	K	L
0.8	96	0.6	0.6	11.9	70	306	0.8	1.3	4	353	11.9
1.2	90	0.6	0.5	92.7	60	337	2.2	3.5	2	0	12.1
-7	71	0.6	0	0	60	282	2.6	4.2	5	353	11.8
-7	70	0.6	0	0	70	277	3.5	5.9	5	353	11.9
-5.6	89	0.6	0.2	0.2	70	196	2.2	3.3	5	353	11.9
-4.9	84	0.6	1	94.6	60	30	0.9	2.2	3	0	12
-4.9	74	0.6	0.1	92.8	70	7	2	2.7	3	0	12
-4.9	78	0.6	0.5	93.3	70	74	2.1	3.5	3	0	11.8
-4.9	78	0.6	0.3	93.7	60	245	2.1	3.7	3	0	12.1
-4.7	83	0.6	0.8	95.4	60	110	3.7	4.5	3	0	12
-4.5	82	0.6	0.3	95.7	60	105	2.9	6.1	3	0	12
-4	82	0.6	0.3	96	60	201	1.2	2.4	3	0	11.9
-2.9	88	0.6	0.1	0.2	70	172	5.9	10.8	5	354	11.9
-0.8	77	0.6	0	97.4	70	0	1	4.4	75	0	12
-0.4	89	0.6	0.2	0.7	70	182	3.7	6.2	35	354	11.9
0	91	0.6	0.2	0.9	70	190	3.6	7.9	46	354	11.9
0.1	85	1.8	0.5	98.5	60	99	2	3.1	40	0	12
0.5	93	1.2	0.7	1.9	60	194	3.1	5.1	14	354	11.9
0.7	88	1.2	1	100.9	60	109	2.2	4.1	10	1	12
0.8	96	3.6	2.6	4.6	60	198	2.7	4.7	5	354	11.9
0.9	89	2.4	3.2	104.1	60	107	1.4	2.3	3	1	11.9
0.6	96	3	3.1	7.7	60	201	2.3	3.9	4	354	11.9
1.1	89	2.4	3.1	107.2	60	9	0.9	1.9	2	1	12.1
0.5	97	1.8	2	9.7	60	223	1.8	3.2	4	354	11.9
1	89	1.8	3	110.2	60	295	1.4	3	2	1	12
0.6	97	3	2.3	12	60	229	2.4	4.2	4	354	11.9
0.9	89	3.6	2.8	112.9	60	349	1	1.9	2	0	12
0.4	98	1.2	2	14	60	213	1.3	2.6	4	354	11.9
1.3	88	2.4	2.8	115.8	60	285	2	3.9	2	0	12
0.7	98	1.2	1.2	15.3	60	242	2.5	4.5	4	353	11.8
1.2	89	3	2.4	118.2	60	264	2.3	7.3	2	1	12
-4.5	82	0.6	0.2	120.6	70	297	1.4	2.6	3	0	12
-7	86	0.6	1.1	17.7	70	172	2.6	4.1	5	354	11.9
-6.4	82	2.4	1	121.8	70	99	2.1	2.9	3	0	12.1
-6.5	86	1.8	1	122.8	70	103	1.4	5.4	3	0	12.1
-6.1	76	0.6	0	123.5	60	333	2.2	5.4	3	1	11.9
-5.9	87	0.6	0.4	19.4	70	284	3.3	5.3	5	353	11.7
-5.5	84	0.6	0.9	124.4	60	320	1.7	3.8	3	0	12
-5.3	89	1.2	1.2	125.6	60	322	1.6	4.3	3	0	12
-5.8	89	0.6	0.9	126.5	70	25	1.9	5.2	3	1	12
-8.2	83	1.2	0.1	21.6	70	338	6.4	11.5	5	353	11.9
-8.1	82	0.6	0.2	127.1	70	325	2.8	4.9	3	0	11.9
-8.9	81	0.6	0.2	21.9	60	337	6.4	10.4	6	353	11.8

A	B	C	D	E	F	G	H	I	J	K	L
11.8	98	4.2	2.2	28.8	60	210	5.1	9.4	3	354	12
12.4	93	0.6	0.4	131.3	60	265	1.9	5	0	1	12
1.5	86	0.6	1.6	23.5	60	182	2.6	4.2	4	354	11.9
1.7	90	0.6	0	24.2	60	223	4.3	5.9	4	354	11.8
1.6	93	0.6	0.5	24.7	60	239	3.9	8.2	4	354	11.9
1.9	87	1.8	0.4	129.3	60	275	2.2	4.3	3	1	12.1
1.7	90	0.6	1.4	130.7	60	128	3.5	6.3	2	1	12
12.2	98	6.6	2.9	31.7	60	242	4.4	6.6	3	354	11.9
11.8	92	3	1.5	132.7	60	276	1.8	4.3	0	1	12.1
3.1	96	4.2	2.5	45.4	60	4	1.8	2.6	5	354	11.9
3	97	4.2	3.8	49.2	70	58	3.4	4.5	4	354	11.9
2.4	95	6	1.8	41.1	60	9	2.3	3.3	22	354	11.9
2.9	88	1.8	0.5	135.7	60	81	2.2	2.4	16	0	12.1
2.9	73	0.6	0.3	59.4	60	205	6.3	11.2	4	354	11.8
5.2	89	4.8	1.1	60.5	60	202	8	16.3	14	354	11.9
5.2	84	0.6	0.1	137.8	60	107	2.2	9	11	1	12.1
9.3	90	0.6	0.1	0.7	60	202	9.4	15.9	16	354	12
10.2	86	3	1.7	141.1	60	273	3.4	7.7	22	1	12.1
6.9	77	0.6	2.9	144	60	267	4.7	9.1	5	1	12.1
1.2	98	2.4	2	6.4	70	103	3.8	4.9	29	354	11.9
1.2	94	2.4	2.8	166.4	60	105	2.4	3.7	23	1	12
1.9	99	13.8	1.2	1.2	60	10	6.9	8.7	4	354	11.9
2.4	94	10.2	3.5	147.7	70	352	2.1	5.1	1	0	12.1
1.8	99	1.2	5	5	60	5	6.3	10.9	4	353	11.8
0.3	91	0.6	0	1.5	60	128	2.7	3.8	35	354	11.9
0.4	96	1.2	1.5	3	60	133	3.1	3.9	57	354	11.9
0.8	88	1.2	0.5	158.7	60	103	2.8	4.2	70	0	12
-1.7	95	8.4	3.8	3.8	70	12	7.9	12.4	4	354	11.9
-2	94	0.6	0.2	0.2	60	0	5	9.3	5	354	11.9
-0.3	64	0.6	0.1	156.7	60	93	1.4	2.7	15	1	12.1
0.7	97	1.2	0.3	4.3	60	128	1.1	1.6	30	354	11.9
0.8	93	3.6	0.5	163.6	60	109	2.7	3.3	36	0	12
1.3	98	0.6	0.6	9.7	60	112	2	2.5	4	354	11.8
1.3	98	1.2	1.7	8	60	120	2.4	3	25	354	11.8
1.9	94	0.6	3.3	169.7	60	109	2.4	4.2	20	1	12
0.7	93	1.8	3.3	162	70	106	3.2	4.7	38	1	12.1
1.2	98	1.8	1.1	9.1	70	120	1.7	2.2	8	354	11.9
2.3	94	2.4	1.1	170.8	60	110	3.4	5.3	11	1	11.8
1.8	99	6.6	1.9	12.2	70	86	3.7	5.4	4	354	11.9
1.6	99	2.4	1.5	1.5	60	140	1.2	1.5	4	353	11.9
3.2	96	6.6	1.1	177.9	60	112	3	4.3	6	1	12.1
2.3	99	0.6	0.2	1.7	70	145	2.1	2.5	4	354	11.9
5.8	100	3	5	6.9	70	353	0.8	2.2	4	354	11.9

A	B	C	D	E	F	G	H	I	J	K	L
3.8	96	6	1.8	180.3	60	31	0.6	0.9	4	0	12.1
2.1	96	1.2	0.2	14.6	70	103	1.4	2.5	4	354	11.9
3.5	88	0.6	0	183.6	60	104	2.8	4.1	2	0	12
2.2	97	0.6	1.4	16	60	91	2	3.2	4	354	11.8
3.2	92	5.4	1	184.6	60	107	2.8	3.7	2	1	12
2.3	97	1.2	2	18	60	147	2.4	3.1	4	354	11.8
3.6	91	0.6	1.3	185.9	60	102	2.6	5.8	2	1	12
2.3	98	2.4	1.6	19.6	70	106	3.8	5.2	4	354	11.9
3.1	92	3	1.1	187	60	100	3.1	5.4	2	0	12
1.9	98	1.8	1.8	21.4	70	134	0.9	1.6	4	352	11.9
3.2	92	1.8	1.9	189	60	104	3.4	5.4	2	1	12.1
2.1	98	3.6	2.3	23.7	60	50	1.9	2.4	4	354	11.9
2.8	94	2.4	1.9	190.8	60	127	2.1	3.3	2	0	12.1
2.6	99	10.2	5.4	29.2	60	85	5.5	6.9	4	354	11.9
2.7	94	2.4	2.9	193.7	70	149	1.5	2.5	3	1	12.1
2.6	99	4.8	7.7	36.9	60	79	5.6	6.5	4	354	11.9
3	95	1.2	1.9	195.6	60	140	1.8	3.5	2	0	12
2.7	100	3.6	2	42.7	60	68	4.9	5.7	4	353	11.9
3.2	95	1.8	2	199	60	123	1	2.4	2	1	12
2.5	99	2.4	3.9	40.8	70	80	5.6	6.3	4	354	11.9
3.1	95	0.6	1.4	196.9	60	143	1.9	3.8	3	1	12
2.6	100	0.6	1.1	1.1	70	28	4.2	5.4	4	354	11.8
3.2	95	0.6	0.9	199.8	60	155	1.4	2.3	2	0	12
2.6	100	1.2	0.8	1.9	70	46	3.9	5.4	4	354	11.9
3.1	95	1.2	0.3	200.1	60	164	1.4	2.2	2	0	12.1
3.1	94	0.6	0.1	201.4	60	125	1.5	2.4	2	0	12.1
3.4	94	0.6	0.1	201.5	60	148	1.8	3.9	2	0	12.1
2	94	5.4	0.5	4.5	70	111	4.5	6.3	4	354	11.9
2.3	88	1.8	0.1	201.7	60	100	2.1	3.5	2	1	12.1
1.5	96	1.8	1.1	11.9	60	318	2.7	5.2	4	354	11.8
1.9	86	0.6	0.2	203.3	70	269	2.6	5.3	2	1	12.1
2.7	82	0.6	0.1	202.7	60	254	2.5	5	2	0	11.8
1.7	95	0.6	0.2	10.7	60	329	3	7.3	4	354	11.8
1.7	87	0.6	0.1	203.5	60	316	1.5	2.9	2	1	12
1.4	94	0.6	0.4	12.2	60	302	1.9	3.7	4	354	11.8
1.1	85	0.6	0.1	203.8	60	333	1.8	3.6	2	0	12.1
-4	81	1.2	0.2	13	60	335	2.3	3.6	6	353	11.9
2.9	84	1.8	0.9	14.6	60	278	8	12.1	4	353	11.9
2.6	86	0.6	0.1	15.2	60	284	5	7.3	4	354	11.9
3.7	100	1.2	0.4	0.4	60	12	5	7.1	4	354	11.9
6.9	97	0.6	2	207.2	60	135	1.9	2.7	1	1	12.1
4.4	100	0.6	4.4	13	60	118	1.2	2.5	4	354	11.8
8.2	96	9.6	2.4	215.9	60	98	2.4	3.6	0	1	12.1

A	B	C	D	E	F	G	H	I	J	K	L
11	97	0.6	2.2	218.8	60	0	0.4	0.9	61	1	11.8
11.9	92	1.2	1.5	11.3	60	215	2.7	3.6	7	354	12
12	91	0.6	0.2	9.7	60	200	2.9	4.7	3	355	11.9
13	84	0.6	0.2	221.4	60	12	0.8	1.4	0	1	12.1
11.9	90	1.2	3.3	9.5	60	208	2.4	3.9	3	354	11.9
12.5	84	0.6	2	221.2	60	267	1	2.9	0	1	12.1
11.8	74	1.2	0.2	6.2	60	233	2	3.7	2	355	11.9
3.4	94	3	0.7	229.4	60	30	1.1	1.8	1	1	12
3.2	94	4.2	2.7	14.8	60	22	5.4	8.5	10	354	11.9
2.8	88	0.6	0.1	222.4	60	21	3	3.5	3	0	12
3.3	97	39	4.2	4.2	60	21	4.6	6.6	4	354	11.9
3.2	92	4.8	2.6	226.2	60	57	3.4	4	1	1	12
3.3	94	0.6	2.4	228.6	60	55	2.5	3.2	1	0	12.1
0.9	87	0.6	0.2	231.8	60	340	2.2	3.3	2	0	12
5.4	99	1.8	1.3	1.8	60	239	1.4	2.8	4	354	11.8
5.1	93	0.6	0.7	231.4	60	354	2.1	3.3	1	1	12.1
6.9	81	1.2	0.1	2.5	60	155	3.8	8.1	3	354	11.9
14.6	96	3	1.7	5.4	60	205	3.1	4.9	2	355	11.9
13.3	91	0.6	19	24.4	60	207	4.9	8	2	354	11.9
13.7	91	0.6	4.7	238.3	60	113	4	5.6	0	1	12
13.4	91	1.8	3.8	28.2	60	277	2.2	5.5	3	354	11.9
12.8	86	0.6	2.6	240.9	60	17	1	1.6	0	1	12
13	92	0.6	8.6	36.8	60	251	2.8	4.8	3	355	11.9
12.7	96	3.6	10	46.7	60	252	1.6	2.1	3	354	11.9
12.8	86	1.2	2	243.8	60	308	0.8	2.5	0	1	12.1
13	88	10.8	4.5	248.2	60	329	2.2	3.5	0	1	12
11.5	100	9	13.6	69.6	60	162	2.8	3.5	3	354	12
12.4	91	9	8.9	257.2	60	111	2.7	4.2	0	1	12.1
2.1	85	0.6	0.1	73.9	60	289	6	10.7	15	354	11.9
2.3	85	0.6	0.2	260.2	60	279	1	4.3	15	1	11.9
-0.3	80	0.6	0	260.9	60	343	4.1	7.4	2	1	12
12.8	87	8.4	13.9	93.5	60	215	6.9	11.6	2	354	12
12.8	83	4.2	4.1	266.9	60	277	3.8	6.9	0	1	12.1
-2.8	88	1.2	1.1	271.4	60	94	2.1	4.4	2	0	12
-2.6	88	0.6	0.2	269.2	60	107	2.2	4.1	2	0	12
-2.4	87	0.6	0.2	269	60	109	2.7	5.3	2	0	12
-2.5	83	1.2	0.3	267.7	60	111	2.7	4.5	2	0	12
-2.3	86	0.6	0.6	268.3	70	113	2.8	3.9	2	0	12
-3	86	0.6	0	0.4	70	136	9.3	9.7	5	354	11.9
-2.8	83	1.2	0.6	270.3	60	126	1.4	4.2	2	0	12
-2.8	86	0.6	0.2	272.1	60	121	1.5	5.4	12	0	12.1
-4	86	3	0.4	273.6	60	355	2.6	5.1	2	0	11.9
-4.6	89	1.8	1.2	1.2	60	73	9.8	10.7	18	354	11.9

A	B	C	D	E	F	G	H	I	J	K	L
-4	90	0.6	1.7	277	60	19	8.2	10.8	13	0	12
-5.9	80	0.6	0.1	277.1	60	356	3.2	6.3	144	0	11.9
-4.6	84	1.2	0.6	277.7	60	13	8.1	12	55	0	12.1
0.9	70	1.8	0.2	0.2	70	160	4.8	7	4	354	11.9
1.2	74	0.6	1.1	279.1	60	106	2.4	5.2	1	1	11.9
1.2	71	1.8	0.4	279.5	60	105	2.7	4.1	1	1	12
7.1	100	0.6	0.2	0.2	60	150	3.2	3.6	4	354	11.8
9.6	95	0.6	2.2	7.1	60	107	3.7	5.9	1	1	11.9
5.2	73	4.8	5.5	6.4	60	60	5	6.1	4	354	11.9
4.7	73	9.6	2	281.6	60	161	2	3.1	1	1	12
3.9	93	2.4	6	12.4	60	163	3	4.3	4	354	11.9
5	85	1.2	0.2	19.2	60	115	0.5	1.2	1	1	12
5.5	94	2.4	0.3	1.9	60	7	2.9	4.2	9	1	12.1
8	95	3	0.5	4.9	60	108	4.4	5.7	1	1	12
4.7	83	3	0.4	17.5	60	206	1.1	2	2	1	12
4.3	98	17.4	2.7	2.7	60	74	7.5	8.3	4	354	12
4.5	95	2.4	12.5	12.5	60	103	2.7	3.9	1	1	11.8
5.1	49	3.6	1.2	284.6	60	59	2.2	4.7	0	1	12
3.9	91	3.6	0.3	0.3	60	327	5.5	7.1	4	354	11.9
3.6	89	0.6	0.9	285.5	60	23	2.4	3.4	2	1	12.1
3.9	91	3.6	0.3	0.3	60	327	5.5	7.1	4	354	11.9
6.9	100	9.6	0.4	0.4	60	151	4.2	4.7	5	354	12
9.1	95	5.4	2.9	10.3	60	117	2.2	5.2	2	1	12.1
3.8	97	1.8	0.9	1	60	0	0	0	4	0	11.7
3.9	98	1.2	0.1	0.1	60	13	7	10.3	7	354	11.9
4.2	98	8.4	4.3	4.3	60	20	7.2	9.9	29	354	11.9
4.1	94	1.2	0.1	288.3	60	27	2.8	3.8	15	1	12
4.2	99	0.6	2.4	6.7	60	19	5.6	7.8	16	354	11.9
4.2	95	0.6	0.4	288.7	60	17	3	4.3	8	1	12.1
6.5	100	1.8	0.1	10	60	5	3.3	7.7	59	354	11.9
5.1	99	6.6	0.6	13	60	16	5.1	5.9	4	354	11.8
4.7	97	3	1.4	6.9	60	86	6.9	10.2	4	354	11.8
4.8	95	6	5.8	6.8	60	108	3.3	4.8	1	1	12
7.9	95	0.6	0.1	3.9	60	103	2.6	3.5	1	1	12
7.2	100	0.6	0.5	0.5	60	141	3.3	4	4	354	11.9
9.2	95	0.6	0.3	7.4	60	102	3.5	6.5	1	1	12.1
8.3	100	2.4	7.3	7.7	60	146	3.6	4.7	19	354	12
9.3	95	0.6	5.5	15.8	60	102	3.4	5	3	1	12.1
4.7	83	0.6	0	0	60	68	9.9	12.5	116	354	11.9
5.2	99	10.8	3	3	60	76	6.1	7	83	354	12
4.8	96	4.8	5.6	8.9	60	96	2.7	3.3	24	1	12.1
4.9	98	6.6	12.3	28.1	60	111	7.1	8.6	24	354	11.9
4.6	94	1.2	1.7	1.7	60	107	2.8	7	10	1	12

A	B	C	D	E	F	G	H	I	J	K	L
4.6	86	0.6	0.1	0.1	60	0	0	0	4	0	11.8
5.1	86	2.4	1.3	1.4	60	60	5.9	7.7	4	354	11.9
5.1	86	2.4	1.3	1.4	60	60	5.9	7.7	4	354	11.8
4.9	90	6	6.1	6.1	60	68	9.1	11.2	4	354	11.9
4.8	88	1.8	0.3	20.6	60	88	3.5	5.2	1	1	12.1
4.9	90	6	6.1	6.1	60	68	9.1	11.2	4	354	11.9
4.8	88	1.8	0.3	20.6	60	88	3.5	5.2	1	1	12.1
5.2	99	22.2	24.5	27.2	60	87	6.2	7.7	4	354	11.9
4.5	95	12	1.1	1.1	60	112	3.5	5.4	1	1	12
4.5	98	64.2	5.6	5.6	60	0	0	0	4	0	11.9
4.8	95	6	5.8	6.8	60	108	3.3	4.8	1	1	11.9
4.1	99	27	15.8	15.8	60	0	0	0	6	0	11.9
4.4	93	8.4	4.7	4.8	60	106	3.5	7.9	2	1	12.1
5.2	96	5.4	1.1	29.6	60	109	4.5	5.8	28	354	11.9
5.3	98	1.2	2.3	2.3	60	68	7.4	8.2	57	354	11.9
5	96	5.4	2.7	11.6	60	103	2.2	3.3	25	1	11.9
5.4	99	3	12.7	15	60	25	3.9	5.8	17	354	11.9
4.8	96	3	5.8	17.5	60	0	0.3	0.9	18	1	12.1
5.7	97	4.2	1.1	24.1	60	24	2.3	2.9	53	354	11.9
7.1	99	8.4	3.3	27.4	60	117	2.6	4.1	27	354	11.9
6	94	0.6	0.8	21.8	60	133	1.4	3.4	24	1	12.1
6	96	1.2	1	22.8	60	22	1.9	2.6	38	1	12
6.6	100	0.6	0.1	27.7	60	57	1.3	2	50	354	11.9
9	100	6.6	1.1	28.9	60	106	3.2	4.7	37	354	11.9
5.9	96	1.2	1	24	60	30	1.9	2.1	45	1	12.1
7.3	100	3.6	3	31.9	60	81	2.3	2.8	11	354	11.9
5.9	97	9	3.3	27.3	60	17	2	2.5	6	1	11.9
5.2	82	1.2	0	27.6	60	99	0.7	1.5	0	1	12
5.8	90	11.4	2.9	2.9	60	72	8.8	10.5	4	354	11.8
5.3	91	1.8	1.8	30.9	60	99	3.6	5.3	4	1	12
5.2	97	25.8	5.3	5.3	60	68	6.7	7.7	12	354	11.9
5.2	91	7.2	4.7	35.6	60	102	3.3	4.8	12	1	12
5	98	10.8	14	14	60	68	7.2	8.7	27	354	11.9
5	93	6.6	5.3	40.8	60	90	4.1	5.8	29	1	12.1
5	99	7.2	6.5	20.5	60	77	7.7	9.2	86	354	11.9
4.6	95	0.6	2.2	43.1	60	78	5.2	7.2	46	0	12.1
4.6	93	8.4	1.1	46.2	60	32	8	11.6	12	1	11.9
4.6	97	15	0.8	0.8	60	28	12.4	18.8	16	354	11.9
4.6	92	1.2	3.1	49.3	60	33	7.7	10.3	7	1	12.1
4.6	97	15	0.8	0.8	60	28	12.4	18.8	6	354	11.9
4.6	91	1.8	1.5	50.8	60	39	8.7	11.8	2	1	12
4.3	94	7.8	3	3	60	0	0	0	4	0	11.9
4.6	89	1.2	0.7	51.6	60	29	7.9	11	1	1	12

A	B	C	D	E	F	G	H	I	J	K	L
4.3	93	18	12.3	12.4	60	26	9.6	13.3	4	354	11.9
4.6	88	2.4	1.5	1.5	60	26	7.6	11.9	1	1	12
4.3	93	18	12.3	12.4	60	26	9.6	13.3	4	354	11.9
4.6	88	2.4	1.5	1.5	60	26	7.6	11.9	1	1	12.1
4.1	93	28.2	10.4	10.4	60	38	14.9	16.3	4	0	11.9
4.6	89	4.8	5.6	5.6	60	31	7.3	10.8	1	1	12
4	88	3	3	3	60	26	10	12.6	1	1	12.1
3.8	93	3.6	0.3	0.3	60	19	9.8	15.1	4	354	11.9
4	91	7.2	4.7	4.7	60	13	6.9	10.7	1	1	12.1
3.4	95	3.6	0.6	0.6	60	16	15	20.1	4	354	11.9
3.9	90	7.2	6.2	6.2	60	0	8.4	14.2	1	1	12
3.5	96	21.6	2.7	2.7	60	0	0	0	4	0	11.7
3.9	91	1.8	0.3	0.3	60	5	8.3	13	1	1	12.1
3.6	96	4.2	1.1	1.1	60	7	9.3	18.2	4	354	11.9
2.6	88	1.8	0.6	4.3	60	4	7.1	12.4	1	1	12.1
15.4	74	1.2	2.4	8.5	60	334	2.4	3.6	33	2	11.9
6.4	89	0.6	0.4	0.4	60	107	4.2	5.7	41	354	11.8
6	84	2.4	0.9	20.9	60	124	2.6	4.7	33	1	12.1
5.2	71	0.6	0.2	14.2	60	89	1.7	2.3	1	1	12.1
6.1	78	1.8	1.7	27.5	60	64	3.6	4.8	5	354	11.9
5.5	79	6.6	1.9	16.1	60	32	3.6	4.2	2	1	12.1
5.6	89	12.6	11.3	38.8	60	77	6.4	7.5	19	354	11.9
5.2	83	1.2	3.9	20	60	90	3	3.8	12	1	11.9
16.6	82	2.4	1.9	4.3	60	253	5.7	7.7	1	354	11.8
15.8	86	0.6	3.7	25	60	80	0.6	0.8	-2	1	12.1
15.7	90	0.6	11.4	38.7	60	262	3.7	4.8	2	354	11.9
15.6	87	4.2	4.3	32.8	60	262	2.7	5.9	0	1	11.9
15.5	94	4.8	7.2	46	60	258	3.6	5.3	2	355	11.9
15.4	90	4.8	7.6	40.4	60	265	2.2	6.1	0	1	12.1
15.5	96	6.6	5.1	51.1	60	222	3.5	5.6	2	354	11.9
15.2	89	0.6	5.4	45.8	60	275	2.4	6.1	0	1	12
18.7	77	21.6	2	57.2	60	262	8	12.3	127	355	12
18.2	70	0.6	0.3	48.1	60	280	1.6	5.2	149	1	12
16.9	75	0.6	0.7	58.2	60	275	2.9	4.6	1	355	11.9
15.8	95	0.6	1.5	67.3	60	175	2.8	4.9	2	355	11.9
16.6	89	2.4	3.3	61.5	60	156	1.1	2	1	354	11.9
16.3	80	1.2	4.2	52.3	60	102	1.4	2.6	-2	1	12.1
6.9	95	0.6	0.4	0.8	60	60	4.1	7.1	4	354	11.9
6.3	93	0.6	0.1	3.6	60	54	5.4	6.4	1	1	12
15.6	97	4.2	2.5	63.9	60	152	0.8	1.8	2	354	11.9
15.8	87	1.8	2.6	54.9	60	112	2.7	3.6	0	1	12.1
15.3	90	0.6	0.9	68.9	60	196	5.4	8	2	355	11.9
14.4	91	1.8	1.2	71.2	60	194	4.6	8.9	16	354	11.9

A	B	C	D	E	F	G	H	I	J	K	L
14	82	0.6	0.2	59.3	60	67	1.8	5	4	1	12.1
5.1	97	2.4	2.4	2.4	60	0	0	0	28	0	11.9
5	92	6	4.6	5.8	60	22	7.8	10.4	10	1	12.1
5.6	86	13.2	0.4	73.4	60	26	5.4	8.2	89	354	12
5.5	89	3	8.6	8.7	60	22	8	15.4	143	354	11.9
4.8	83	1.8	0.7	60.9	60	18	8.1	10.2	117	1	12
4.9	89	0.6	0.1	0.1	60	43	7.4	10.9	32	1	12
5.2	95	20.4	3.1	3.1	60	44	10.5	14.9	32	354	11.9
4.8	89	0.6	1.2	1.2	60	38	8.5	10.8	24	1	12.1
5.1	97	2.4	2.4	2.4	60	0	0	0	10	0	11.9
5.3	92	2.4	4.4	4.4	60	22	9.5	13.2	8	1	12.1
5.1	97	2.4	2.4	2.4	60	0	0	0	6	0	11.9
5.5	92	6.6	7.1	11.5	60	27	9.1	11.4	3	1	12.1
5.1	97	2.4	2.4	2.4	60	0	0	0	4	0	11.9
6.3	93	5.4	3.6	3.6	60	29	8.6	10.5	1	1	12.1
5.1	97	2.4	2.4	2.4	60	0	0	0	4	0	11.9
6.8	93	6	0.7	0.7	60	0	0	0	1	0	12
6.8	97	6.6	0.7	0.7	60	39	9.7	11.4	4	354	11.9
6.8	93	6	0.7	0.7	60	0	0	0	1	0	12.1
6.9	98	1.2	1	1.8	60	61	4.5	8.9	44	354	11.9
6.9	98	1.2	1	1.8	60	61	4.5	8.9	21	354	11.9
6.9	98	1.2	1	1.8	60	61	4.5	8.9	21	354	11.8
6.4	94	0.6	0.1	4.2	60	64	7	8.2	13	1	11.9
6.6	95	0.6	2.5	6.6	60	83	6.6	8.9	24	1	12.1
7.2	95	3	0.2	7.8	60	7	4.1	4.7	40	1	12.1
7	100	0.6	1.1	1.1	60	39	5.5	8.6	7	354	11.9
6.6	97	0.6	0.1	8	60	22	4.5	5.2	3	1	12.1
6.9	97	1.8	4.1	12.1	60	46	3.9	4.6	1	1	11.8
7.9	95	1.2	0.2	12.4	60	42	1.2	1.7	1	1	12.1
8.7	100	0.6	3.5	9.4	60	46	3.1	6.2	3	354	11.9
10	100	19.8	7.4	16.8	60	232	4	8.2	3	353	11.9
9.9	98	4.8	4.6	17.9	60	255	1.4	3.2	0	1	12
12.2	100	2.4	2.5	19.4	60	220	0.6	1.1	3	354	11.9
14.3	95	1.8	2.8	22.1	60	161	1.4	2	2	354	11.8
9.5	82	0.6	0.8	0.9	60	240	6	10	660	354	12
9.8	79	3	0.6	25	60	265	3.3	8	79	1	12.1
8.9	85	1.2	0	25.5	60	275	2.8	9.3	0	1	12.1
5.1	87	1.2	5.5	5.5	60	269	4.7	7.1	4	354	11.9
4.9	85	1.2	1.7	7.2	60	268	5.4	9.8	8	353	11.8
4.9	85	1.2	1.7	7.2	60	268	5.4	9.8	28	353	11.9
4.8	83	0.6	2	2	60	267	8.6	12.5	69	354	11.9
6.9	86	2.4	1	3.7	60	106	7.5	11.5	109	354	11.9
6.5	71	2.4	0.6	4.8	60	287	1.9	4.2	3	354	11.9

A	B	C	D	E	F	G	H	I	J	K	L
3.2	53	1.8	0.3	27.3	60	54	1.3	1.6	0	1	12
5.2	56	0.6	0.2	4.1	60	19	0.9	1.3	3	354	11.9
4.6	56	1.2	1.1	30.7	60	110	0.5	1.2	1	1	12.1
3.8	58	7.2	2.5	2.5	60	96	4.2	7.2	4	354	11.8
4.2	69	0.6	1.8	32.5	60	106	2.8	4.4	1	1	12
5.1	72	1.8	0.3	4.9	60	102	2.9	5	12	353	11.8
4.8	71	1.2	0.2	35.3	60	111	2.5	4.5	9	1	12.1
9.5	100	2.4	3.5	3.5	60	20	7	8.5	3	0	11.8
9	96	1.8	2.4	8.2	60	69	2.7	3.8	0	1	12.1
9	81	1.2	3.9	9.7	60	114	1.3	2.4	193	354	11.8
9.6	85	1.8	0.8	10.5	60	113	1	1.6	100	354	11.9
9	90	0.6	0.8	41.1	60	0	0.8	1.5	3	1	11.9
9.3	95	8.4	5.3	46.4	60	13	5.9	9.7	0	1	12.1
10.2	100	7.8	22.7	34.9	60	30	7.4	11.9	3	354	11.9
8.5	96	6.6	5.8	5.8	60	25	2.2	3.7	0	1	12.1
17.1	64	7.2	1	9.6	60	341	2.4	3.7	-3	1	12.1
17.1	88	6	4.1	4.1	60	279	2.4	4.3	-2	1	12.1
5.8	88	12	4.4	4.4	60	33	6.6	7.8	70	354	11.9
6.2	86	4.8	0.7	5.2	60	42	4.4	5.8	49	1	12
7.6	90	2.4	0.1	6.2	60	24	8.1	11.2	69	1	12
11.4	95	13.2	9.1	9.1	60	124	5.1	6.4	45	354	11.9
10.4	93	6.6	0.7	0.7	60	107	3.2	5.8	38	1	11.9
13.1	95	6	1.1	18.5	60	350	0.6	1.8	22	354	11.9
12.8	91	1.8	0.3	7.2	60	77	0.8	1.2	21	1	12
12.4	98	4.2	1.9	29.6	60	9	4.3	6.9	3	354	12
11.7	95	7.2	2.8	16	60	3	4	5.9	0	1	11.9
15	65	3.6	0.3	12.7	60	254	4.8	9.3	2	355	12
14.4	80	0.6	0.4	13.9	60	256	5	7.5	1	354	11.9
14.2	82	1.2	0.3	5.4	60	278	1.9	3.1	91	359	12.1
14.1	82	0.6	0.3	16.8	60	258	2.7	3.5	290	355	11.9
13.7	72	0.6	0	16.8	60	257	2.9	4.4	297	354	12
13.2	90	0.6	3.2	21.7	60	22	1.7	3.3	67	354	12
12	87	1.2	0.2	11.5	60	82	0.8	1.5	0	1	12
12.7	97	0.6	0.7	24.3	60	10	3.3	5.5	2	354	11.9
12.7	98	4.8	3.4	27.7	60	14	5.3	8.4	2	354	11.9
11.5	94	1.8	1.2	13.1	60	39	1.7	2.7	0	1	12.1
12.5	98	7.8	9.7	39.3	60	357	3.4	5.8	3	354	11.9
11.7	94	1.2	3.8	19.7	60	343	1.5	3.4	0	1	12.1
10.7	97	6	1.9	1.9	60	8	7	12.2	38	354	11.9
12.3	91	0.6	6.9	22	60	62	5.8	9.7	26	354	12
11.6	93	7.2	6	33.1	60	18	2.3	3.5	6	1	12.1
12.8	87	3	0.1	5.3	60	46	1.9	3	27	354	11.9
12.3	75	3.6	0.1	20.8	60	63	0.8	1.5	14	1	11.9

A	B	C	D	E	F	G	H	I	J	K	L
12.8	90	22.8	9.8	15.1	60	69	5.2	7.4	26	354	11.9
11.9	89	4.8	6.3	27.1	60	13	1.7	2.3	17	1	12
16	59	0.6	0.7	34.5	60	269	3	5.4	0	354	11.9
18.7	71	0.6	0	39.6	60	114	0.4	1.1	-3	1	12
20.7	68	1.2	1.5	41.2	60	239	0.9	3.1	71	2	12.1
20.7	62	4.8	1.3	42.5	60	10	4.1	7	4	1	12
21.2	88	3.6	6.4	55.4	60	271	3.5	7.9	15	355	11.9
20.8	91	15	3.6	47	60	287	1.5	3.6	3	1	12.1
19.7	90	3.6	4	59.5	60	23	2	3.4	7	354	11.9
20.1	89	3	4.4	51.4	60	347	0.8	2.7	2	1	12.1
18.2	79	0.6	0.6	70.2	60	272	7.4	11.7	32	354	11.9
18	73	0.6	0.2	54.1	60	270	3	7.8	19	1	12.1
18.5	86	3	1.5	71.7	60	243	3.7	6	75	354	11.9
19.8	54	1.8	0.1	53.6	60	345	2.7	3.8	-5	2	12.1
18.6	81	1.8	5.7	68.1	60	231	6.7	12.6	1	354	12
22.8	72	0.6	0	55.5	60	273	0.8	2.1	110	2	12
20.6	89	0.6	1.4	74	60	18	0.9	1.2	0	355	12
19.4	92	0.6	2.1	58.3	60	96	0.7	1.5	-2	1	12
21.3	91	0.6	0.4	74.3	60	153	0.9	2	0	355	12
17.6	54	0.6	0.4	75	60	270	4.1	7.5	241	355	11.9
14.9	59	1.2	0.1	81.8	60	349	2.1	3	21	354	11.9
14.7	83	0.6	0.1	82.8	60	118	1	1.8	66	354	11.9
24	59	1.8	0.2	60.3	60	0	0.7	2.6	86	2	11.9
24.1	64	6.6	0.1	85.8	60	218	4.8	10	426	355	11.9
19.4	100	15	17.8	103.6	60	333	1.5	2.3	1	355	12
17.8	96	9	7.4	69.8	60	52	0.5	0.8	-2	1	12.1
16.4	70	12	4	113.5	60	222	3.4	6.6	79	355	11.9
16.4	71	2.4	2.6	75.5	60	273	0.5	1.3	73	1	12.1
19.6	67	3.6	5	126.2	60	194	5.3	9.7	176	354	11.9
20.5	63	18	0.6	126.9	60	216	4	6.5	0	355	12
20.2	74	1.8	1.7	128.6	60	197	2.8	6.4	0	355	12
18.5	80	0.6	0.5	78	60	350	2.3	3.3	-3	1	12.1
21.5	85	14.4	2	135.1	60	286	2.5	5.2	5	355	12
19.9	73	6.6	1.3	81.7	60	5	2.6	4.2	0	1	12.1
20.5	92	3.6	2	137.1	60	154	1.4	2.3	0	355	11.9
19.2	78	2.4	1.3	83	60	212	0.4	0.7	-3	1	12
22.9	87	64.8	6.1	146	60	247	6	8.9	32	355	12
22.1	70	1.2	0.3	86.4	60	282	1.5	3.8	-4	1	12
22.3	82	3.6	1.5	149	60	231	2.9	4.9	0	354	11.9
22.4	76	1.2	1.7	88.1	60	275	1.6	4.6	-4	1	12
22.1	89	12	6.9	155.8	60	9	5	10.1	0	355	11.9
19.4	91	10.8	5.6	93.7	60	357	2.7	4.7	-3	1	12.1
19.9	96	4.8	11.4	167.2	60	8	4.1	7.4	5	354	11.9

A	B	C	D	E	F	G	H	I	J	K	L
18.9	93	9	7.6	101.3	60	11	1.6	2.8	2	1	12
19.8	98	13.8	12	179.3	60	180	3.9	7.1	30	355	11.9
19.4	86	0.6	8.1	109.4	60	90	2.2	4.8	19	1	12.1
23.4	78	0.6	0	109.8	60	90	1	2.2	86	1	12
24.5	94	0.6	1.7	184.4	60	184	1.6	3.4	0	355	11.9
24.2	89	0.6	0.8	110.9	60	0	0	0.6	-4	2	12
20.8	84	3.6	4.8	116.6	60	340	1.8	3.7	-2	1	12
20.3	100	31.8	34.7	220.4	60	245	2.9	3.6	9	355	11.9
21.9	96	4.2	9.2	235.1	60	199	2.6	5.9	73	355	11.9
22.2	80	2.4	4.7	128.9	60	237	0.7	3.5	38	1	12.1
25	92	1.2	0	236.2	60	238	3.7	5.7	0	355	11.9
24.8	84	0.6	0.1	129.4	60	268	1.6	3.9	-5	1	12.1
22.1	65	0.6	0	129.7	60	356	1.8	2.8	585	3	12.1
20.6	98	2.4	12.1	248.8	60	6	2.5	3.4	0	354	12
20.9	93	3	0.9	254	60	171	0.8	1.8	1	354	12
21.3	89	1.2	1.9	139.3	60	335	1.6	3.7	-3	3	12
20.8	100	13.8	16.9	271	60	81	1.7	2.4	2	354	12
20.3	94	9	6.9	146.2	60	125	0.6	1	-2	3	12.1
21.1	100	4.2	7	277.9	60	29	2	2.9	2	354	11.9
20.4	93	8.4	9.8	156.1	60	16	3.4	5.3	0	2	12
23.7	92	1.8	1.5	284	60	204	0.6	1.3	38	355	12
23.8	97	23.4	11.7	295.7	60	5	1.3	1.7	0	355	11.9
22.3	94	2.4	1.7	169	60	0	0.4	1.2	-4	3	12.1
23.5	98	5.4	5.9	301.6	60	154	1.2	4.1	0	355	11.9
22.3	93	0.6	4.2	173.1	60	90	0.9	2.9	-4	3	11.9
23.1	97	3	2.6	304.2	60	146	0.8	2.2	0	355	12
22.9	91	7.8	1.9	175.1	60	117	1.1	2.1	-4	3	12.1
23.8	98	3	7	311.2	60	199	2.9	5	0	354	12
22.7	93	15	7.2	182.3	60	119	1.1	2.3	-4	3	12
22.9	99	26.4	12.4	323.7	60	224	4	7.5	0	354	11.9
22	89	2.4	3.2	185.5	60	302	1	2.1	-3	3	12.1
22.9	99	3	8.5	332.2	60	223	2.3	4.4	0	355	11.9
21.9	92	3	5.9	191.4	60	245	1.5	3.5	-4	3	12
23.1	100	7.2	4.1	338	60	256	3.5	5.5	0	354	11.9
22.3	93	3.6	3.3	196.2	60	250	1.5	3.4	-4	3	12
22.7	99	22.2	13.1	351.2	60	216	4.2	7.4	0	354	11.9
22.1	86	6	5.1	201.3	60	311	2.6	5.3	-3	3	12
21.9	95	0.6	4	369.7	60	258	5.3	7.6	7	354	11.9
21.7	90	1.2	0.8	205.9	60	336	2	4	1	3	12
17.9	71	0.6	0	206.8	60	110	1.3	2.5	39	3	12.1
23	84	0.6	0	374.2	60	0	0	0.3	0	355	11.9
23.4	90	0.6	1.8	376	60	181	2	4	0	355	11.9
23.2	83	1.8	0.5	207.4	60	122	0.5	1	-4	3	12.1

A	B	C	D	E	F	G	H	I	J	K	L
19.6	69	0.6	4.4	212.2	60	97	1.1	2.3	111	3	12.2
22.7	84	16.8	1	378.8	60	259	2.4	3.7	7	355	12
22.4	91	5.4	5.1	384	60	50	3.3	8	0	355	11.9
20.6	83	13.8	3.5	217	60	190	1.4	3.8	-3	3	12.1
19.3	92	13.2	12.3	419.2	60	249	4.7	7.1	1	354	11.9
19.2	84	0.6	4.2	221.1	60	269	1.8	4.2	-2	2	12
20.2	85	0.6	0	223.2	60	99	1	1.9	0	3	12.1
22.9	75	0.6	0.2	223.4	60	111	1.8	3.1	37	3	12.1
23.6	84	0.6	1.8	225.2	60	81	0.4	0.8	75	3	11.9
22	95	22.8	4.1	429.6	60	239	3.5	5.4	0	354	12
21.7	88	1.2	2.1	227.8	60	263	2.2	5.5	11	3	12.1
19.4	74	0.6	0	230.9	60	90	1.3	2	775	4	12.1
19.5	71	0.6	0	231	60	105	1.2	2.6	749	4	12.1
23.1	61	0.6	0.1	231.1	60	107	0.8	1.5	9	3	12
20.8	98	6	19.1	457.1	60	234	3.6	5.9	0	355	11.9
21.8	94	4.8	4.6	461.9	60	261	4	7	0	355	11.9
21.8	89	1.8	1.1	239	60	284	1.9	3.9	-3	3	12.1
16.9	57	0.6	0	465.9	60	132	2.6	4	0	354	11.8
18	80	0.6	2	242.6	60	1	0.9	1.4	-2	3	12.2
18.3	85	0.6	2.2	244.8	60	344	0.5	1.2	-3	3	12.1
17.7	77	3.6	0.7	246.1	60	99	0.9	1.5	-3	3	12.1
17.5	76	0.6	0.1	467.2	60	266	1.6	2.2	0	354	11.9
19.3	87	57	6.5	474	60	285	2.5	4.6	10	355	12
13.7	82	4.2	8	489.2	60	279	1.8	4.1	2	354	11.9
22.2	76	3	0	492.7	60	323	3.4	7.7	44	354	12
12	78	0.6	0.8	253	60	58	2.4	3.2	49	3	12.1
16.6	78	1.8	0.6	494.7	60	76	3.5	4	6	355	11.9
19.1	92	1.8	2.5	498.3	60	59	0.9	2.2	1	354	11.9
18.8	87	0.6	0.5	256.5	60	84	1	1.6	-3	3	12
18.4	89	2.4	3.5	501.8	60	74	0.5	0.9	1	354	11.9
18.9	90	1.2	3.5	260	60	116	0.5	0.9	-3	3	12
18.1	93	1.8	3.3	505.1	60	65	1.1	1.5	1	354	11.9
18.1	87	0.6	1.1	261.1	60	0	0	0.3	-3	3	12.1
19.5	96	2.4	1.1	508.1	60	113	0.6	1	50	354	11.9
4.5	78	5	5	5	5	299	2.2	4.5	-4	5	12
16.2	85	0.6	1.2	515.7	60	27	3.8	6.4	1	354	11.9
15.6	83	3.6	1.2	265.3	60	110	1.9	2.7	-3	3	12
19.6	87	4.2	0.7	267.6	60	279	0.8	4.2	0	3	12.1
19.3	92	4.8	1.2	522.1	60	196	2.8	6.6	1	354	11.9
19.1	93	0.6	1.6	523.7	60	184	3	5.5	1	355	11.9
18.7	84	0.6	0.5	268.8	60	107	2	2.8	-3	3	12.2
17.9	73	1.2	0.3	269.6	60	92	1.2	2.8	-3	3	12.1
17.3	85	1.2	3.5	529.9	60	273	9	13.2	10	355	11.9

A	B	C	D	E	F	G	H	I	J	K	L
16.3	94	33	14	544	60	252	6.8	11.1	5	354	11.9
16	90	14.4	6.8	276.5	60	275	1.6	5.5	1	3	12.1
15.8	90	2.4	21.7	565.7	60	226	4.9	8.7	12	355	11.9
15.6	84	4.2	5.5	282	60	282	2.3	6	26	2	11.8
16.4	85	2.4	0.2	568	60	258	7.4	11.6	1	354	11.9
15.8	81	4.8	0.5	568.6	60	36	6	10.1	4	354	11.9
15.1	97	19.8	18.7	587.3	60	355	3.8	6.7	2	354	11.9
14.6	86	13.8	5.6	288.7	60	19	4.4	6.8	0	2	12
15.5	99	3.6	6.3	593.6	60	26	7	10.8	2	354	11.9
16	92	1.8	3.9	597.5	60	58	4.8	9.3	1	354	11.9
15.1	89	1.8	1.7	295.4	60	34	5.4	7.1	-2	2	12.1
16.2	100	34.8	22.9	620.5	60	72	10.6	13.8	1	354	12
15.8	92	10.2	8.6	303.9	60	94	5.6	8.6	-2	2	12.1
16.5	100	7.8	31.9	652.4	60	93	8.1	12.2	1	354	12
16.2	92	7.2	9.5	313.4	60	92	5.8	8.5	-2	3	12.1
16.6	100	12	6.8	659.2	60	87	9.5	13.1	1	355	12
16.2	89	1.8	2.8	316.2	60	95	7	9.5	-2	2	12.1
16.2	90	9	1.8	317.9	60	101	5.6	10	-2	3	12.1
16.4	100	12.6	10.1	672.8	60	99	7.5	13	1	354	12
16.1	91	1.8	2	320	60	96	7.5	11.3	-2	2	12.1
15.7	100	1.8	9.4	682.2	60	89	7.3	13.9	1	355	11.9
15.9	89	1.2	1.2	321.2	60	99	7.9	11.6	-2	3	12.1
15.4	100	7.2	12.8	695	60	102	6.6	11.8	2	354	11.9
15.1	90	12	8.7	329.9	60	108	8.2	12.4	-2	2	12.1
14.2	100	14.4	13.7	708.8	60	95	8.9	16.5	2	354	11.9
14.6	91	4.2	11.5	341.4	60	104	7.8	11.8	-2	3	12.1
14.4	88	0.6	0.4	341.8	60	101	6.1	9.7	-2	3	12.1
14.1	100	15.6	19.5	730.2	60	77	8.4	10.3	2	354	12
13.8	100	0.6	14.8	745.1	60	71	8.4	10.4	13	354	12
13.7	89	6.6	5.5	350.9	60	76	6.7	8.1	14	3	11.9
16.1	90	6	5.2	357	60	61	1.2	1.6	-2	3	12
16.8	95	31.8	1.3	758.2	60	71	3.3	3.8	1	354	11.9
15.8	94	0.6	1.7	365.5	60	110	1.7	3	-2	3	12.1
16.5	95	1.8	0.7	758.9	60	120	1.4	2.1	3	354	11.9
16.5	97	3.6	11	770	60	165	2.5	3.9	5	354	11.8
16.1	94	3	4.9	372.6	60	119	1.1	2	-2	3	12.1
16	92	8.4	8.1	380.8	60	104	2	3.2	0	3	12.1
16.2	98	1.8	14	795	60	154	1.4	4.4	14	354	11.9
16.1	94	3.6	5.7	386.5	60	236	1.6	3.4	22	3	12
16.3	97	7.8	17.5	812.5	60	176	2.7	4.8	19	354	11.9
16.2	96	19.2	8.4	820.9	60	146	1.7	2.7	36	354	11.9
16.3	92	5.4	4.8	394.1	60	126	2.6	3.8	31	3	12.1
17	98	3.6	8.1	829	60	313	1.6	3.3	3	354	11.9

A	B	C	D	E	F	G	H	I	J	K	L
16.5	90	9.6	3.5	397.6	60	118	1.9	3.2	0	3	12
17	100	1.2	9.8	838.9	60	153	1	3	23	354	11.9
16.9	100	6	2.4	841.3	60	137	1.2	1.8	22	354	11.9
17.1	93	6.6	1	402.8	60	112	2.8	3.8	29	3	12.2
16.9	95	2.4	1.2	404	60	105	2.1	2.9	15	3	12.1
17	95	3.6	2	406	60	84	0.9	1.7	0	3	12.1
18.7	94	0.6	0.3	412.2	60	98	2.8	4.6	-3	3	12
18.3	98	0.6	0.4	851	60	155	2.7	5.9	1	354	12
18.2	89	3.6	2.9	415.1	60	100	3.2	5.6	-3	3	12
17.6	100	6.6	34.9	886	60	14	3.9	5.6	1	354	11.9
17.2	94	2.4	6.6	421.7	60	134	1.1	1.6	-3	3	12.1
17.5	93	4.2	2.5	428.3	60	95	1.1	4.8	-3	3	12.1
18	100	15	20.5	914.1	60	169	2.5	3.9	1	354	12
17.9	95	8.4	9.2	437.5	60	105	2.6	4.5	-3	2	12.1
19.2	94	0.6	1.4	927.5	60	241	5.4	10.3	1	354	11.8
19.3	95	10.8	2.3	929.7	60	247	5.4	8.8	1	354	11.9
18.7	89	9	0.8	445.4	60	298	1.9	3.9	-3	3	12
18.6	94	0.6	3	932.8	60	266	7.7	11.2	1	354	11.8
15.7	90	0.6	0.8	447.8	60	337	2.2	5.6	-2	3	12.1
12.2	99	19.2	3.4	940.2	60	33	5.1	9.4	2	354	12
11.7	91	4.2	0.9	451	60	18	4.7	5.4	0	2	12.1
12.9	100	1.8	5.8	946	60	22	5	7.2	2	354	11.9
12.3	93	1.2	4.1	455.2	60	10	3.8	4.6	0	2	12.1
12.7	96	1.2	2.1	948.1	60	13	6	9.4	2	354	11.9
12.4	97	2.4	1.4	949.5	60	10	3.2	6.4	2	354	11.9
11.7	92	0.6	0.6	456.1	60	4	3.8	5.4	0	2	12.1
12.5	97	4.8	1.4	950.9	60	11	5.2	8.2	2	354	11.9
11.6	92	1.8	0.9	457	60	4	3.6	5.3	0	2	12.1
12.1	96	9	4.9	955.8	60	10	5.1	9.3	2	354	12
11.4	92	1.8	1.6	458.7	60	2	2.8	4.8	0	2	12.1
11.9	100	4.2	3.7	959.5	60	7	5.2	9.5	3	354	11.9
11.5	92	1.8	0.8	459.5	60	353	3.8	5.7	0	2	12.1
11.4	100	2.4	7.7	967.1	60	12	6.6	9.9	3	354	11.9
11.4	93	4.2	3.4	462.9	60	355	4.5	6.1	0	2	11.8
11.4	100	2.4	7.7	967.1	60	12	6.6	9.9	3	354	11.9
11.2	93	8.4	6.5	469.4	60	349	3.3	5.5	0	2	12.1
12.2	100	0.6	7.3	7.3	60	7	7.3	13.5	3	354	12
11.5	93	0.6	5.6	475	60	352	4.6	6.7	0	2	12
12.4	100	3	3	10.4	60	13	8.4	12.2	3	355	11.9
12.4	97	3.6	4.9	15.2	60	19	6.1	9.7	7	354	11.9
12.2	92	3	1.2	478	60	13	4.1	6.1	3	2	12.1
12.3	95	1.8	5.1	20.3	60	21	6.7	10.7	21	354	11.9
11.5	89	1.2	1.8	479.8	60	13	5.5	6.7	24	2	11.9

A	B	C	D	E	F	G	H	I	J	K	L
12.1	94	2.4	3.5	23.9	60	17	7	10.8	42	354	11.9
11.7	86	0.6	0.6	480.4	60	14	4.4	6.3	29	2	12.1
12.3	95	7.8	6.5	30.3	60	0	4.2	6.9	20	354	11.9
11.8	89	0.6	1.4	481.8	60	14	5.3	6.8	18	2	12.1
12.1	93	4.8	7.4	37.7	60	14	6.1	10.3	30	354	11.9
10.5	81	3.6	4.4	486.2	60	1	4.4	6.3	24	2	12
11	89	13.8	5.4	5.4	60	350	4.5	7.5	34	354	11.9
10.6	87	2.4	3.9	490.1	60	330	1.8	3.8	25	2	12
9.8	86	3	0.9	492.7	60	273	1.5	3.5	0	2	12.1
9.6	89	3	2.5	495.2	60	290	1.8	3.4	0	2	12.1
8.8	94	2.4	0.3	10.8	60	266	4.6	7	3	354	11.8
8.5	94	0.6	0.7	11.5	60	271	4.5	7.3	3	354	11.9
8.5	95	0.6	0.3	11.8	60	253	3.2	5.4	3	354	11.9
8.5	89	0.6	0.2	497.8	60	266	2.3	4.9	0	2	12.1
8.2	87	3	1	498.8	60	275	2.7	4.9	0	3	11.9
7.4	91	21	4.3	4.3	60	262	7	11.2	19	354	11.9
7.8	89	3	2.4	501.3	60	277	1.1	2.3	4	2	12.1
6.4	92	29.4	1.7	1.7	60	0	0	0	34	0	11.9
7.8	90	1.2	2.9	504.2	60	328	1.5	3.2	25	2	11.9
6.6	93	3	1.4	1.4	60	268	5.3	7.9	49	354	11.8
7.8	90	2.4	2.6	2.6	60	299	1.7	3.7	30	2	12.1
6.2	93	3	4.4	5.8	60	269	6.6	9.9	50	354	11.8
7.4	90	0.6	1.8	4.5	60	1	0.4	2.7	36	2	12
6.2	94	10.8	0.4	0.4	60	273	5.5	7.6	116	354	11.9
7.5	89	3	2.2	6.6	60	272	1.9	3.4	59	2	12.1
6.7	94	3	6.1	6.5	60	271	4.6	8.4	86	354	11.9
7.6	89	4.8	1.5	8.1	60	323	1.7	3.2	52	2	12
7.1	92	0.6	3	3	60	270	3.9	7.1	90	354	11.9
7.2	92	1.2	1.4	4.4	60	271	6.1	9.7	49	354	11.9
7.3	91	0.6	1.4	5.8	60	273	4.4	8.2	52	354	11.8
7.3	90	10.8	1.4	7.2	60	269	6.5	9.4	28	354	11.9
7.2	84	0.6	0.3	11.3	60	326	1.1	2.1	10	2	12.1
7	87	9	2.2	9.4	60	283	3.3	8.5	4	354	11.9
6.7	80	0.6	0.7	12	60	325	1	4.6	1	2	12.1
6.7	88	0.6	2.5	11.9	60	289	2.3	4.9	3	354	11.9
-0.4	50	0.6	0.1	12.5	60	309	1.2	3.2	85	2	11.9
4.6	80	1.8	1	13	60	177	2.4	3.6	19	354	11.9
4.6	71	1.2	0.2	12.7	60	98	3.1	6.1	11	2	12
4.8	74	1.2	0.3	13.3	60	169	4.2	6.9	7	354	11.9
5.2	78	2.4	3.6	16.3	60	111	2.9	4.3	4	2	12
5.2	83	3	3.3	16.6	60	167	3.6	5.6	4	354	11.9
5.2	79	2.4	3.5	19.8	60	91	2.3	4.5	0	2	12.1
5.5	83	8.4	3.2	23	60	105	2.7	4.5	0	2	12.1

A	B	C	D	E	F	G	H	I	J	K	L
6	93	2.4	3.7	22.3	60	151	2.2	4.8	3	353	11.9
6	93	2.4	3.7	22.3	60	151	2.2	4.8	3	353	11.9
6.3	87	1.2	1.8	27.5	60	115	2	4.6	0	2	12
9.9	95	0.6	0	24	60	2	2.4	3.9	23	354	11.9
9.6	97	0.6	0.1	24.3	60	31	2.4	4.7	3	354	11.9
7	95	3.6	2.2	27.4	60	118	4.3	5.2	48	354	11.9
7.3	94	3.6	0.2	27.6	60	123	3.1	4.6	44	354	11.9

Bibliography

[1] Adnan Aijaz and A Hamid Aghvami. Cognitive machine-to-machine communications for internet-of-things: A protocol stack perspective. *IEEE Internet of Things Journal*, 2(2):103–112, 2015.

[2] Kamal M Ali and Michael J Pazzani. Error reduction through learning multiple descriptions. *Machine Learning*, 24(3):173–202, 1996.

[3] Gustavo EAPA Batista, Ana LC Bazzan, and Maria Carolina Monard. Balancing training data for automated annotation of keywords: a case study. In *WOB*, pages 10–18, 2003.

[4] Shen Bin, Liu Yuan, and Wang Xiaoyi. Research on data mining models for the internet of things. In *2010 International Conference on Image Analysis and Signal Processing*, pages 127–132. IEEE, 2010.

[5] Nitesh V Chawla, Kevin W Bowyer, Lawrence O Hall, and W Philip Kegelmeyer. Smote: synthetic minority over-sampling technique. *Journal of Artificial Intelligence Research*, 16:321–357, 2002.

[6] Wassim Derguech, Eanna Bruke, and Edward Curry. An autonomic approach to real-time predictive analytics using open data and internet of things. In *2014 IEEE 11th Intl Conf on Ubiquitous Intelligence and Computing and 2014 IEEE 11th Intl Conf on Autonomic and Trusted Computing and 2014 IEEE 14th Intl Conf on Scalable Computing and Communications and Its Associated Workshops*, pages 204–211. IEEE, 2014.

[7] Cao-Tri Do, Ahlame Douzal-Chouakria, Sylvain Marié, and Michèle Rombaut. Multiple metric learning for large margin knn classification of time series. In *2015 23rd European Signal Processing Conference (EUSIPCO)*, pages 2346–2350. IEEE, 2015.

[8] Vassilis Foteinos, Dimitris Kelaidonis, George Poulios, Panagiotis Vlacheas, Vera Stavroulaki, and Panagiotis Demestichas. Cognitive management for the internet of things: A framework for enabling autonomous applications. *IEEE Vehicular Technology Magazine*, 8(4):90–99, 2013.

[9] Peter Hart. The condensed nearest neighbor rule (corresp.). *IEEE Transactions on Information Theory*, 14(3):515–516, 1968.

[10] Simon Haykin. Cognitive radio: brain-empowered wireless communications. *IEEE Journal on Selected Areas in Communications*, 23(2):201–220, 2005.

[11] ShaoHua Hu. Research on data fusion of the internet of things. In *2015 International Conference on Logistics, Informatics and Service Sciences (LISS)*, pages 1–5. IEEE, 2015.

[12] Vikramaditya Jakkula and Diane Cook. Outlier detection in smart environment structured power datasets. In *2010 Sixth International Conference on Intelligent Environments*, pages 29–33. IEEE, 2010.

[13] Nathalie Japkowicz et al. Learning from imbalanced data sets: a comparison of various strategies. In *AAAI Workshop on Learning from Imbalanced Data Sets*, volume 68, pages 10–15. Menlo Park, CA, 2000.

[14] Michael I Jordan and Robert A Jacobs. Hierarchical mixtures of experts and the em algorithm. *Neural Computation*, 6(2):181–214, 1994.

[15] Mohamed Amine Kafi, Yacine Challal, Djamel Djenouri, Messaoud Doudou, Abdelmadjid Bouabdallah, and Nadjib Badache. A study of wireless sensor networks for urban traffic monitoring: applications and architectures. *Procedia Computer Science*, 19:617–626, 2013.

[16] Muhammad Aamir Khan, Aunsia Khan, Muhammad Nasir Khan, and Sajid Anwar. A novel learning method to classify data streams in the internet of things. In *2014 National Software Engineering Conference*, pages 61–66. IEEE, 2014.

[17] Miroslav Kubat, Robert C Holte, and Stan Matwin. Machine learning for the detection of oil spills in satellite radar images. *Machine Learning*, 30(2-3):195–215, 1998.

[18] Miroslav Kubat, Stan Matwin, et al. Addressing the curse of imbalanced training sets: one-sided selection. In *ICML*, volume 97, pages 179–186. Nashville, TN, 1997.

[19] Jorma Laurikkala. Improving identification of difficult small classes by balancing class distribution. *Artificial Intelligence in Medicine*, pages 63–66, 2001.

[20] Charles X Ling and Chenghui Li. Data mining for direct marketing: Problems and solutions. In *KDD*, volume 98, pages 73–79, 1998.

[21] Xiaolei Ma, Yao-Jan Wu, Yinhai Wang, Feng Chen, and Jianfeng Liu. Mining smart card data for transit riders' travel patterns. *Transportation Research Part C: Emerging Technologies*, 36:1–12, 2013.

[22] Mohammad Saeid Mahdavinejad, Mohammadreza Rezvan, Mohammadamin Barekatain, Peyman Adibi, Payam Barnaghi, and Amit P Sheth. Machine learning for internet of things data analysis: A survey. *Digital Communications and Networks*, 4(3):161–175, 2018.

[23] Aaisha Makkar and Neeraj Kumar. Cognitive spammer: a framework for pagerank analysis with split by over-sampling and train by under-fitting. *Future Generation Computer Systems*, 90:381–404, 2019.

[24] Joseph Mitola and Gerald Q Maguire. Cognitive radio: making software radios more personal. *IEEE personal communications*, 6(4):13–18, 1999.

[25] Peng Ni, Chunhong Zhang, and Yang Ji. A hybrid method for short-term sensor data forecasting in internet of things. In *2014 11th International Conference on Fuzzy Systems and Knowledge Discovery (FSKD)*, pages 369–373. IEEE, 2014.

[26] Michael Peter Perrone. *Improving regression estimation: Averaging methods for variance reduction with extensions to general convex measure optimization.* PhD thesis, Citeseer, 1993.

[27] Yongrui Qin, Quan Z Sheng, Nickolas JG Falkner, Schahram Dustdar, Hua Wang, and Athanasios V Vasilakos. When things matter: A survey on data-centric internet of things. *Journal of Network and Computer Applications*, 64:137–153, 2016.

[28] Deborah E Rosen and Elizabeth Purinton. Website design: Viewing the web as a cognitive landscape. *Journal of Business Research*, 57(7):787–794, 2004.

[29] Swaytha Sasidharan, Andrey Somov, Abdur Rahim Biswas, and Raffaele Giaffreda. Cognitive management framework for internet of things:—a prototype implementation. In *Internet of Things (WF-IoT), 2014 IEEE World Forum on*, pages 538–543. IEEE, 2014.

[30] Amit Sheth. Transforming big data into smart data: Deriving value via harnessing volume, variety, and velocity using semantic techniques and technologies. In *2014 IEEE 30th International Conference on Data Engineering*, page 2. IEEE, 2014.

[31] Madhu Shukla, YP Kosta, and Prashant Chauhan. Analysis and evaluation of outlier detection algorithms in data streams. In *2015 International Conference on Computer, Communication and Control (IC4)*, pages 1–8. IEEE, 2015.

[32] AH Schistad Solberg and Rune Solberg. A large-scale evaluation of features for automatic detection of oil spills in ers sar images. In *Geoscience and Remote Sensing Symposium, 1996. IGARSS'96. Remote Sensing for a Sustainable Future., International*, volume 3, pages 1484–1486. IEEE, 1996.

[33] Ivan Tomek. Two modifications of cnn. *IEEE Trans. Systems, Man and Cybernetics*, 6:769–772, 1976.

[34] Durga Toshniwal et al. Clustering techniques for streaming data-a survey. In *2013 3rd IEEE International Advance Computing Conference (IACC)*, pages 951–956. IEEE, 2013.

[35] Panagiotis Vlacheas, Raffaele Giaffreda, Vera Stavroulaki, Dimitris Kelaidonis, Vassilis Foteinos, George Poulios, Panagiotis Demestichas, Andrey Somov, Abdur Rahim Biswas, and Klaus Moessner. Enabling smart cities through a cognitive management framework for the internet of things. *IEEE Communications Magazine*, 51(6):102–111, 2013.

[36] David H Wolpert. Stacked generalization. *Neural Networks*, 5(2):241–259, 1992.

[37] Shenyong Xiao, Han Yu, Yanan Wu, Zijun Peng, and Yin Zhang. Self-evolving trading strategy integrating internet of things and big data. *IEEE Internet of Things Journal*, 2017.

[38] Mingchuan Zhang, Haixia Zhao, Ruijuan Zheng, Qingtao Wu, and Wangyang Wei. Cognitive internet of things: concepts and application example. *International Journal of Computer Science Issues*, 9(6):151–158, 2012.

Index